AdS/CFT Correspondence
in Condensed Matter

AdS/CFT Correspondence
in Condensed Matter

Antonio Sergio Teixeira Pires

Universidade Federal de Minas Gerais, Departamento de Física,
Belo Horizonte, MG, Brazil

Morgan & Claypool Publishers

Rights & Permissions
To obtain permission to re-use copyrighted material from Morgan & Claypool Publishers, please contact info@morganclaypool.com.

ISBN   978-1-627-05309-9 (ebook)
ISBN   978-1-627-05308-2 (print)

DOI   10.1088/978-1-627-05309-9

Version: 20140601

IOP Concise Physics
ISSN 2053-2571 (online)
ISSN 2054-7307 (print)

A Morgan & Claypool publication as part of IOP Concise Physics

Morgan & Claypool Publishers, 40 Oak Drive, San Rafael, CA, 94903, USA

Contents

Preface

The goal of this text is to introduce, in a very elementary way, the concept of anti-de Sitter/Conformal Field Theory (AdS/CFT) correspondence to condensed matter physicists. This theory relates a gravity theory in a $(d+1)$-dimensional anti-de Sitter space time to a strongly coupled d-dimensional quantum field theory living on its boundary. The AdS/CFT correspondence can be used to study finite temperature real time processes, such as response functions and dynamics far from equilibrium in quantum critical points in condensed matter systems. Computation of these quantities is reduced to solving classical gravitational equations in one higher dimension than the original theory.

Very little in this text is my original contribution. It has been collected mainly from the research literature. Some time ago, I published a short version in the arXiv, and received a lot of e-mails with questions and suggestions that I have incorporated in the present text.

Here I intend to introduce AdS/CFT correspondence to condensed matter physicists, presenting just the general ideas. The interested reader can consult the references to get more information. I have presented only relevant materials, omitting the detail. There are a lot of papers and very good reviews about the subject, but most of these use a language that demands some knowledge of the subject. Here, I have tried to use a more accessible language and therefore some sections may seem to be very simple to the reader. In chapter 2, I introduce quantum phase transitions to physicists with no knowledge in the subject. In chapter 3, I give a brief overview of general relativity theory, so that the reader can perform calculations without knowing the subject, just following the methodology. In chapter 4, it is shown that choosing a metric in a manifold with an extra dimension and with all the symmetries of field theory at the critical point, leads to a metric for a space which is the AdS space. A simple example of a scalar field is then presented, however the theory is at zero temperature. To work at finite temperature a black hole is introduced into the theory in section 4.3. In condensed matter physics we are generally interested in the presence of an electric or magnetic field; this is treated in section 4.4. In chapters 5 and 6, I present a brief introduction to dynamics. Holographic superconductors and fermions are studied in chapter 7, and finally in chapter 8, I present my conclusions.

It is impossible to thank all the people who have contributed to this book, but I would especially like to thank Tobias Wenger for some relevant information.

Author biography

Antonio Sergio Teixeira Pires

Antonio Sergio Teixeira Pires (born 18 November 1948) is a Professor of Physics in the Department of Physics at the Universidade Federal de Minas Gerais, Belo Horizonte, Brazil. He received his PhD in Physics from University of California in Santa Barbara in 1976. He works in techniques of quantum field theory applied to condensed matter. He is a member of the Brazilian Academy of Science, was an Editor of the Brazilian Journal of Physics and currently is a member of the Advisory Board of the Journal of Condensed Matter Physics.

AdS/CFT Correspondence in Condensed Matter

Antonio Sergio Teixeira Pires

Chapter 1

Introduction

1.1 Gauge theories

In the literature concerning anti-de Sitter/Conformal Field Theory (AdS/CFT) correspondence, gauge theory is typically used (the word gauge is used only for historical reason). Since this is a book for the condensed matter physicist, some of whom may be unfamiliar with the theory, I will present a brief overview here. It is not my aim to teach the subject here but only to explain its meaning. Although all matter particles (leptons and quarks) are described by the Dirac equation, for simplicity I consider the Klein–Gordon equation which describes a spin 0 particle. The same procedure can be used in the Dirac equation, but it is more cumbersome because we then have to deal with matrices. The Klein–Gordon equation

$$\nabla^2 \psi - \frac{1}{c^2} \frac{\partial^2 \psi}{\partial t^2} = m^2 c^2 \psi, \tag{1.1}$$

can be derived from the Lagrangian

$$L = -\left(\frac{\partial \psi^*}{\partial x} \frac{\partial \psi}{\partial x} + \frac{\partial \psi^*}{\partial y} \frac{\partial \psi}{\partial y} + \frac{\partial \psi^*}{\partial z} \frac{\partial \psi}{\partial z} - \frac{1}{c^2} \frac{\partial \psi^*}{\partial t} \frac{\partial \psi}{\partial t} + m^2 \psi^* \psi \right), \tag{1.2}$$

where I have taken $\hbar = 1$. It is easy to see that the transformation

$$\psi(\vec{r}, t) \rightarrow e^{i\alpha} \psi(\vec{r}, t), \tag{1.3}$$

(where α is a real constant) leaves the Lagrangian invariant. We refer to this transformation as a 'global phase transformation'. Let us now consider the case where α is an arbitrary real function of position and time. The transformations at different points of space time are independent of one another, and we have a local phase transformation given by

$$\psi\left(\vec{r}, t\right) \rightarrow e^{i\alpha\left(\vec{r},t\right)}\psi\left(\vec{r}, t\right). \tag{1.4}$$

Calculating the derivative of ψ we obtain

$$\frac{\partial \psi}{\partial x_{\mu}} = \left(\frac{\partial \psi}{\partial x_{\mu}} + i\frac{\partial \alpha}{\partial x_{\mu}}\right)e^{i\alpha} \tag{1.5}$$

where $x_{\mu} = x, y, z, t$. We see that the new function is not a solution for the original equation (1.1) because of the derivative of α. So far there is nothing to worry about. The next step is to demand that the Lagrangian (1.2) (and of course, equation (1.1)) be invariant under the local phase transformation. There is no need for this to be true, but if we believe that it happens, let us go ahead. To cancel the term originating from the derivative of α, we need to add a new term in the Lagrangian. To do that, we introduce a vector field $A_{\mu}(x_{\mu})$ (called the gauge field) that changes according to the rule

$$A_{\mu} \rightarrow A_{\mu} + \frac{1}{q}\frac{\partial \alpha}{\partial x_{\mu}}, \tag{1.6}$$

(where for now q is just a constant) and we replace the ordinary derivative in the Lagrangian by

$$\frac{\partial \psi}{\partial x_{\mu}} \rightarrow D_{\mu}\psi \equiv \left(\frac{\partial}{\partial x_{\mu}} + iqA_{\mu}\right)\psi. \tag{1.7}$$

We now have

$$D_{\mu}\psi \rightarrow e^{i\alpha}\left(D_{\mu}\psi\right), \tag{1.8}$$

and the equation written in terms of D_{μ} (called the covariant derivative) is invariant under the local phase transformation. We have to pay a price for introducing a new field that couples to ψ. For consistency we need to add to the Lagrangian an extra term that describes the field $A_{\mu}(x_{\mu})$. To define a closed dynamical system this term must involve $\partial_{\mu}A_{\mu}$ quadratically. The only gauge Lorentz scalar of this type is given by

$$L = -\frac{1}{16\pi}F^{\mu\nu}F_{\mu\nu}, \qquad \text{where} \quad F_{\mu\nu} = \partial_{\mu}A_{\nu} - \partial_{\nu}A_{\mu}, \tag{1.9}$$

and where the sum over repeated indices is implied. The factor $1/16\pi$ is just a question of convention. To keep the invariance the field $A_{\mu}(x_{\mu})$ must be massless. Equation (1.9) is just the Maxwell Lagrangian. We see therefore that $A_{\mu}(x_{\mu})$ is the electromagnetic potential and q the electric charge that works as a coupling constant. In a sense, the existence and form of the Maxwell equation have been derived by demanding that local phase transformations leave the original Lagrangian invariant.

We can consider the local phase transformation in equation (1.3) as a multiplication of ψ by a unitary matrix $U = e^{i\alpha(x_\nu)}$. The group of all such matrices is denoted by $U(1)$ and therefore the symmetry involved is called $U(1)$ gauge invariance.

Now let us consider a more general situation where we have two complex fields ψ_1 and ψ_2, written as a two-component column vector

$$\psi = \begin{pmatrix} \psi_1 \\ \psi_2 \end{pmatrix}. \tag{1.10}$$

Before going ahead (for the benefit of the reader not familiar with the theme) let us consider a column matrix written as

$$\begin{pmatrix} z_1 \\ z_2 \end{pmatrix}, \tag{1.11}$$

where z_1 and z_2 are complex numbers that represent the components of a vector. The modulus of the vector is invariant under a rotation given by a unitary matrix with determinant equal to 1. This matrix is called $SU(2)$. We have only three independent real parameters and we can see this by writing the matrix as

$$U = \begin{pmatrix} a & b \\ c & d \end{pmatrix}. \tag{1.12}$$

Demanding that $U^+ = U^{-1}$ and $\det U = 1$, we find

$$U = \begin{pmatrix} a & b \\ -b^* & a^* \end{pmatrix} \qquad |a|^2 + |b|^2 = 1, \tag{1.13}$$

which shows that we have three independent parameters. For the case of N dimensions we have a rotation matrix $SU(N)$ with $N^2 - 1$ independent real parameters.

Let us go back to the field ψ in equation (1.10). We see that a more general transformation is

$$\psi \to S\psi, \tag{1.14}$$

where S is a matrix function of x_μ. This generalization was proposed by Yang and Mills in 1954 and given the names *Yang and Mills theory* and *Yang and Mills fields*.

As before, the Lagrangian is not invariant under this transformation. The derivative of ψ leads to an extra term given by

$$\frac{\partial \psi}{\partial x_\mu} \to S\frac{\partial \psi}{\partial x_\mu} + \left(\frac{\partial S}{\partial x_\mu}\right)\psi. \tag{1.15}$$

Following the same procedure we define a covariant derivative

$$D_\mu = \frac{\partial}{\partial x_\mu} + iq\vec{\tau}.\vec{A}_\mu, \tag{1.16}$$

where now $\vec{\tau}$ are the Pauli matrices and we have three massless gauge fields $\vec{A}_\mu = \left(A_\mu^1, A_\mu^2, A_\mu^3\right)$ that describe particles with spin ½ and mass zero. The three fields are associated with the three independent parameters mentioned after equation (1.13). As such, the above formalism is of no use: there are no pairs of particles with zero mass that could be described by this theory. A modified theory, starting with the electron and neutrino with zero masses and then coupling with the Higgs field to give mass to the electron and the gauge particles (the bosons W^+, W^-, Z^0) has been used to study the weak interaction. It was the necessity of this coupling to the Higgs field to preserve the concept of gauge symmetry that made the Higgs boson so important.

This becomes more interesting when we look at $SU(3)$. Quantum chromo-dynamics describes quark interactions. In this theory each type of quark (called flavors) comes in three colors—red, blue and green. The three colors of a given flavor have the same mass. Thus a triplet of colors written as

$$\psi = \begin{pmatrix} \psi_r \\ \psi_b \\ \psi_g \end{pmatrix}, \tag{1.17}$$

is invariant under a $SU(3)$ rotation that transforms one quark (with the same flavor) to another. We now have eight independent parameters in the matrix $SU(3)$ that lead to eight massless gauge fields, which we associate with the gluons. A calculation similar to the one that led to the Maxwell equation (although much more complicated since now we have eight fields) can be performed to give the equations of chromodynamics.

The description of leptons and quarks and the gauge bosons (photon, W^+, W^-, Z^0 and the eight gluons) using the $U(1)$, $SU(2)$ and $SU(3)$ gauge transformations constitute what we call the standard model of particle physics. Several extensions of this model have been proposed. In one called *supersymmetry* (*SUSY*), every known elementary particle has a supersymmetric partner (called superpartner) which is like it in all respects except for its spin. The superpartners of fermions have spin 0 and are named by adding the prefix 's' to the name of the fermions (for instance, selectron, squark), while the superpartners of the bosons have spin ½ and are named by adding the suffix 'ino' to the root of the normal name (for instance, photino, gluino). If supersymmetry were exact, a particle and its superparticle would have the same mass, but this is not what happens, since superparticles have not been detected. If the theory is correct, the symmetry has been broken in some way.

String theory is an approach to the study of fundamental interactions and particles based on different principles than the ones used in the standard model. It tries to describe all the interactions in nature, including gravity. It supposes that the fundamental particles are not point-like objects, having a mathematical dimension of zero, but are string-like, one-dimensional objects, vibrating in higher dimensions. Each vibration mode corresponds to a particle and, what is more interesting; one of these modes can be associated to the graviton (the quantum particle of the grav-itational field). In 1976 it was realized that supersymmetry could be incorporated in string theory, forming what was called superstring theory in 10 dimensions. To get rid of the extra dimensions, they can be rolled up to a small size. Because of the very small size of the strings (on the order of 10^{-33} cm), they cannot be resolved experimentally (just to give an idea, the LHC can probe scales of about 10^{-19} cm).

The theory is purely conjectural and there is as yet no evidence to show that it has anything to do with the physics of elementary particles.

I will not go into any further detail here since the above material is sufficient for the purpose of this book.

1.2 Gauge gravity duality

Anti-de Sitter/Conformal Field Theory (AdS/CFT) correspondence [1–3] has become very important in higher energy theoretical physics over the last ten years, and hundreds of papers have been published about the subject. The meaning of these words will be explained later. One reason for this interest is that this correspondence can be used to understand strongly interacting field theories by mapping them to classical gravity. The main idea is as follows: large N gauge theories (which are strongly coupled conformal field theories) in d space time dimensions are mapped to a classical gravitational theory in $(d+1)$ space time dimensions which is asymptotically anti-de Sitter. In other words, the field theory lies in a flat space time in d dimensions, and this can be viewed as the boundary of a 'bulk' space with anti-de Sitter geometry in one extra dimension where weakly coupled gravitational physics is at work (figure 1.1). In fact, this is a strong–weak coupling duality: when one theory is coupled weakly, the dual description involves strong coupling, and vice versa. But what is N? Consider a set of n non-interacting fields φ_i, with equal masses and differing only in a quantum number that we call *color*. We introduce a vector field $\phi = (\varphi_1,...,\varphi_n)^T$. The Lagrangian describing ϕ is invariant under the transformation $\phi \to \phi' = S\phi$, where S is a unitary $N \times N$ matrix. This is a generalization of what happens with quarks. A given quark, as was mentioned earlier, can exist in three colors: red, blue and green. In this case (as was mentioned in section 1.1) $N = 3$.

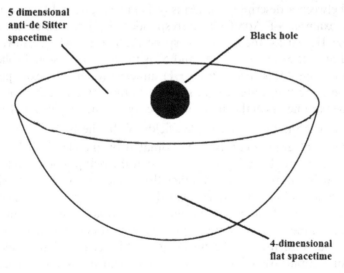

Figure 1.1. Field theory lies within the boundary. The bulk is asymptotically AdS. The meaning of the black hole will be explained later.

Gauge theories with large N are just mathematical constructions; they are generalizations of the theory describing quarks for the case of not three different colors, but nearly an infinite number of colors (large N). In condensed matter it is not clear what large N means. Unlike the large N limit of vector models, large N gauge theory remains strongly coupled.

Different CFTs will correspond to theories of gravity with different field contents and different bulk actions, e.g. different values of the coupling constants in the bulk. AdS/CFT correspondence is an example of a more general technique called 'the holographic principle', which was motivated by the study of the thermodynamic properties of a black hole. (The mapping has been referred to as holography, making a parallel with holograms which contain all the information about the shape of a three-dimensional object in two dimensions.) As is well known, the entropy of a black hole is proportional to the area of its event horizon. Thus, the number of degrees of freedom needed to describe a quantum black hole scales with its area, not its volume. We can interpret this result saying that the physics inside the black hole is mapped onto its horizon, some kind of 'hologram'. The number of degrees of freedom of the d-dimensional (boundary) theory equals those of the $(d + 1)$-dimensional (bulk) theory.

Arguments show that some quantum field theories (QFTs) are secretly quantum theories of gravity and therefore we can use these to compute observations of the QFT when the gravity theory is classical, as found in [4]. The number of theories that can be studied using the correspondence is still small, and it does not include any theory that describes a known physical system. However, it is hoped that these theories can be used to capture essential features of theories realized in nature [1]. The formalism involves not just a different mathematical model, but a different type of setting, and is usually formulated using mathematical tools that are unfamiliar to the majority of condensed matter physicists. A G Green [5] has pointed out that gauge gravity duality is, in some ways, similar to a familiar feature of classical electromagnetism—the duality between a description in terms of field strength and lines of force flux.

The first example of AdS/CFT correspondence [1] presented an equivalence between type IIB string theory in the space $AdS_5 \times S^5$ (where AdS_5 is the five-dimensional anti-de Sitter space time and S^5 is the five-dimensional sphere) and the $N = 4$ $SU(N)$ Yang–Mills theory in $(3 + 1)$ dimensions. In this example, the parameters on the gravity side are the AdS radius L, the string length l_s and the Planck length l_p. The parameters in the dual gauge theory are the so called 't Hooft coupling $\lambda = (L/l_s)^4$ and the number of complex fields N. In the limit $\lambda \to \infty$, $N \to \infty$, the quantum theory of gravity can be approximated by the classical theory. Large L means weak curvature. However we remark that the real physical models have only $N = 1$ and 2. It is still not known whether this conjecture works for other models, however the assumption that it does has led to an impressive set of results.

For a theory to admit a holographic description it must admit a large N expansion and in the large N limit most of the operators in the theory must acquire large anomalous dimensions (the meaning of this term will be explained later). However, for the purpose of the present text we can just say that the field theory should be strongly interacting [6].

Recently, several authors have claimed that some phenomena in condensed matter systems, such as quantum phase transitions, also provide candidates for the use of AdS/CFT [7–35]. This began in 2007, with Sachdev, Son, Hartnoll and others, and since then the number of people working on this subject has increased. As has been pointed out by Zaanen *et al* [31], the techniques used by condensed matter theorists are rather antiquated. They amount largely to the perturbative and mean field theories, with some other techniques such as the Luttinger liquids of one dimension and Chern–Simons topological field theory. AdS/CFT correspondence seems to be a new mathematical machine that could lead to new progress in physics. However, this remains a dream at present, since the ideas are still very fresh.

The best evidence for duality comes from string theory, but most of the details are irrelevant for applications in condensed matter. Therefore, I am not going to justify the application of the formalism but just accept that it works. A motivation for the correspondence without using string theory is presented in [4].

There is a substantial literature of excellent reviews written by professionals (see, for instance [4, 7–9, 22, 24, 31]). Reference [31] in particular gives an excellent very detailed presentation of the subject.

References

[1] Maldacena J M 1998 The large N limit of superconformal field theories and supergravity *Adv. Theor. Math. Phys.* **2** 231
Klebanov I R and Maldacena J M 2009 Solving quantum field theories via curved space-times *Physics Today* January 28
Johnson C V and Steiberg P 2010 What black holes teach about strongly coupled particles *Physics Today* May 29
Nastase H Introduction to AdS-CFT arXiv:0712.06889 [hep-th]
Aharony O, Gubser S S, Maldacena J M, Ooguri H and Oz Y 2000 Large N field theories, string theory and gravity *Phys. Rept.* **323** 183
[2] Gubser S S, Klebanov I R and Polyakov A M 1998 Gauge theory correlators from non-critical string theory *Phys. Lett.* B **428** 105
[3] Witten E 1998 Anti-de Sitter space and holography *Adv. Theor. Mat. Phys.* **2** 253
[4] McGreevy J Holographic duality with a view toward many-body physics arXiv:0909.05182 [hep-th]
[5] Green A G An introduction to gauge gravity duality and its application in condensed matter arXiv:1304.5908 [cond-mat]
[6] Hartnoll S A Horizons, holography and condensed matter arXiv:1106.4324 [hep-th]
[7] Herzog C P, Kovtun P, Sachdev S and Son D T 2007 Quantum critical transport, duality, and M-Theory *Phys. Rev.* D **75** 085020
[8] Hartnoll S A and Kovtun P 2007 Hall conductivity from dyonic black holes *Phys. Rev.* D **76** 066001
[9] Hartnoll S A, Kovtung P K, Muller M and Sachdev S 2007 Theory of the Nernst effect near quantum phase transitions in condensed matter, and in dyonic black holes *Phys. Rev.* B **76** 144502

[10] Hartnoll S A and Herzog C P 2007 Ohm's law at strong coupling: S duality and cyclotron resonance *Phys. Rev.* D **76** 106012

[11] Hartnoll S A and Herzog C P 2008 Impure AdS/CFT *Phys. Rev.* D **77** 106009

[12] Faulkner T, Liu H, McGreevy J and Vegh D 2011 Emergent quantum criticality, Fermi surface, and AdS_2 *Phys. Rev.* D **83** 125002

[13] Cubrovic M, Zaanen J and Schalm K 2009 String theory, quantum phase transitions and the emergent Fermi-liquid *Science* **325** 439
Cubrovic M, Zaanen J and Schalm K Fermions and the AdS/CFT correspondence: quantum phase transitions and the emergent Fermi-liquid arXiv:0904.1993 [hep-th]

[14] Hartnoll S A, Herzog C P and Horowitz G T 2008 Holographic superconductors *JHEP* **0812** 015

[15] Sachdev S and Muller M 2009 Quantum criticality and black holes *J. Phys.: Condens. Matter* **21** 164216
Sachdev S and Keimer B 2011 Quantum Criticality *Physics Today* **64** 29

[16] Sachdev S Finite temperature dissipation and transport near quantum critical points arXiv:0910.1139 [cond-mat]
Sachdev S 2011 The landscape of the Hubbard model *TASI* 2010 *String Theory and its Applications: From meV to Planck Scale* ed M Dine, T Banks and S Sachdev (Singapore: World Scientific) arXiv:1012.0299 [hep-th]

[17] Sachdev S 2011 Strange metals and Ads/CFT correspondence *J. Stat. Mech.* **1011** 11022

[18] Sachdev S 2012 What can gauge-gravity duality teach us about condensed matter physics *Annu. Rev. Cond. Matt. Phys.* **3** 9

[19] Krempa W W and Sachdev S The quasi-normal modes of quantum criticality arXiv: 1210.4166 [cond-mat]

[20] Jensen K Semi-holographic quantum criticality arXiv:1108.0421 [hep-th]

[21] Faulkner T, Iqbal N, Liu H, McGreevy J and Vegh D From black holes to strange metals arXiv:1003.1728 [hep-th]

[22] Hartnoll S A 2009 Lectures on holographic methods for condensed matter physics *Class. Quantum Grav.* **26** 224002

[23] Iqbal N and Liu H 2009 Universality of the hydrodynamic limit in AdS/CFT and the membrane paradigm *Phys. Rev.* D **79** 025023

[24] Herzog C P 2009 Lectures on holographic superfluidity and superconductivity *J. Phys. A: Math. Gen.* **42** 343001

[25] Hartnoll S A Quantum critical dynamics from black holes arXiv:0909.3553 [cond-mat.str-el]

[26] Sachdev S 2011 Condensed matter and AdS/CFT *Lecture Notes in Physics* **828** 273

[27] Pires A S T AdS/CFT correspondence in condensed matter arXiv:1006.5838 [cond-mat.str-el]

[28] Horowitz G T 2011 Surprising connections between general relativity and condensed matter *Class. Quantum Grav.* **28** 114008

[29] Gubser S S and Rocha F D 2010 Peculiar properties of a charged dilatonic black hole in AdS_5 *Phys. Rev.* D **81** 046001

[30] Charmousis C, Gouteraux B, Kim B S, Kiritsis E and Meyer R 2010 Effective holographic theories for low-temperature condensed matter systems *JHEP* **1011** 151

[31] Zaanen J, Sun Y W, Liu Y and Schalm K The AdS/CFT manual for plumbers and electricians www.lorentz.leidenuniv.nl/~kschalm/papers/adscmtreview.pdf

[32] Ren J Quantum critical system from AdS/CFT www.princeton.edu/physics/.../jie-Ren-Thesis.pdf

[33] Rey S-J 2009 String theory on thin semiconductors: holographic realization of Fermi points and surfaces *Prog. Theor. Phys.* Supplement **177** 128

[34] Krempa W W, Sorensen E S and Sachdev S The dynamics of quantum criticality: quantum Monte Carlo and holography arXiv:1309.2941 [cond-mat.str-ele]

[35] Zaanen J 2013 Holographic duality: stealing dimensions from metals *Nat. Phys.* **9** 609

Chapter 2

Quantum phase transitions

In recent years, experimental physicists have discovered a number of exotic states of matter, including superconductors, superfluids, topological conductors and strange metals. As pointed out by Sachdev [1], these states of matter arise from an unimaginably complex web of quantum entanglement among the electrons—so complex that theorists studying these materials have been at a loss to describe them.

Close to the critical point in a quantum phase transition (QPT), the electrons no longer behave independently or even in pairs but become entangled. In this region, correlations show power-law dependences upon separation in space and time that are the emergent result of complicated many body correlations in quantum theory.

The holographic technique has been applied to two broad classes of condensed matter systems, which Sachdev [2] calls 'conformal quantum matter' and 'compressible quantum matter'. The first class includes models with quantum critical points. The second class is composed of systems which have a non-zero and finite compressibility at $T = 0$, such as Fermi liquids and superfluids. Of interest here are the exotic compressible states (the so called non-Fermi liquids) exhibiting strange metal behavior in a variety of compounds including high temperature superconductors. These systems will be considered in chapter 7.

QPTs take place at a temperature of absolute zero, where crossing the phase boundary means that the quantum ground state of the system changes in some fundamental way. This is accomplished by changing not the temperature, but some parameter, let us say g, like pressure, chemical composition or magnetic field in the Hamiltonian of the system [3, 4]. As pointed out by Sachdev [3], a traditional analysis of many body systems would begin from either a weak-coupling Hamiltonian, and then build in interactions among the nearly free excitations, or from a strong-coupling limit, where the local interactions can be taken into account in a precise way, but their propagation is not fully understood. On the other hand, a quantum critical point starts from an intermediate coupling regime and provides

2-1

another perspective on the physics of the system. In any Hamiltonian that admits a QPT there must be two non-commuting operators that describe two competing ordering tendencies. The effects of quantum criticality, different from the classical case, might be observed at high temperatures. This is, in many cases, the signature of how a quantum critical point might extend to high temperatures. Near a quantum critical point g_c, we can express all physical quantities in terms of the deviation from criticality, $\delta = (g - g_c)$. Scaling theory establishes that the correlation length ξ diverges at the transition point as $\xi \approx |\delta|^{-\nu}$, where ν is the correlation length exponent. Quantum fluctuations occur in time, and so we introduce a new correlation 'length' ξ_t which represents the time scale over which the system fluctuates coherently. We have the relation $\xi_t \approx \xi^z$, where z is called the dynamical exponent, and is a measure of the asymmetry between spatial and temporal fluctuations [4]. For more information about QPTs, I recommend the excellent book by Sachdev [3] and his web page.

2.1 Quantum Ising model

The best example to present the ideas of QPTs is the quantum Ising model with a transverse magnetic field described by the Hamiltonian

$$H = -gJ\sum_i \sigma_i^x - J\sum_{\langle i,j \rangle} \sigma_i^z\sigma_j^z, \qquad (2.1)$$

where $J > 0$ is the exchange constant and gJ (with $g > 0$) represents a magnetic field along the x direction. The coupling parameter g is responsible for the QPT. At a first approximation this Hamiltonian can describe the compound $LiHoF_4$. The two states at site i, with eingenvalues $\sigma_i^z = \pm 1$, represent spins oriented up $|\uparrow\rangle_i$ or down $|\downarrow\rangle_i$. When $g = 0$ we have the classical Ising model, the system is an eigenstate of σ^z, and the ground state is given by

$$|\uparrow\rangle = \prod_i |\uparrow\rangle_i,$$

or

$$|\downarrow\rangle = \prod_i |\downarrow\rangle_i, \qquad (2.2)$$

where $|\uparrow\rangle_i$, $|\downarrow\rangle_i$ are eigenstates of σ_i^z. A simple calculation gives

$$\langle 0|\sigma_i^z\sigma_j^z|0\rangle = N_0^2 \qquad \langle 0|\sigma_i^z|0\rangle = \pm N_0. \qquad (2.3)$$

That is, we have long range order with a spontaneous magnetization along the z direction. The term σ^x induces quantum mechanical tunneling that flips the orientation of the spin on a site.

When $g \gg 1$, the first term in (2.1) dominates, and the ground state is given by

$$|0\rangle = \prod_i |\rightarrow\rangle_i, \qquad (2.4)$$

where

$$|\rightarrow\rangle_i = \frac{1}{\sqrt{2}}\left(|\uparrow\rangle_i + |\downarrow\rangle_i\right) \qquad |\leftarrow\rangle_i = \frac{1}{\sqrt{2}}\left(|\uparrow\rangle_i - |\downarrow\rangle_i\right) \qquad (2.5)$$

are eigenstates of σ_i^x with eigenvalues ±1. In this limit we have $\langle 0|\sigma_i^z\sigma_j^z|0\rangle = \delta_{ij}$ when $g \rightarrow \infty$. Perturbative corrections in $1/g$ give

$$\langle 0|\sigma_i^z\sigma_j^z|0\rangle \approx \exp\left(-|x_i - x_j|/\xi\right), \qquad (2.6)$$

showing an exponential decay of the correlation function, and therefore an absence of long range order.

The phase diagram at $T = 0$ can be represented as

$|\uparrow\uparrow\uparrow\uparrow\uparrow\uparrow \ldots\ldots\rangle$ $\qquad\qquad\qquad\qquad$ $|\rightarrow\rightarrow\rightarrow\rightarrow \ldots\ldots\rangle$

0	g_c	g

As pointed out by Sachdev [3], it is not possible for states that obey (2.3) and (2.6) to transform into each other analytically as we vary g. There must exist a critical value $g = g_c$ at which the correlation functions change from one type to the other. This is the point where the QPT takes place. In one dimension $g_c = 1$ [4]. At g_c the ground state is complicated, but we know that it exhibits scale invariance.

2.2 XY model

One model where QPT can also be well understood is the two-dimensional aniso-tropic quantum XY model described by the following Hamiltonian [5, 6]:

$$H = J \sum_{\langle n,m \rangle} \left(S_n^x S_m^x + S_n^y S_m^y\right) + D \sum_n \left(S_n^z\right)^2, \qquad (2.7)$$

where $\langle n,m \rangle$ represents the sum over nearest neighbors on the sites, n, of a regular lattice and $0 \leqslant D < \infty$, where D represents an xy easy-plane single-ion anisotropy. Due to the form of the single-ion anisotropy we should have $S > \frac{1}{2}$ and so we take $S = 1$. For D less than a critical value D_C, the system has a thermal phase transition at a temperature T_{BKT}, the so called Berezinskii–Kosterlitz–Thouless temperature. This is another reason why this model is so interesting to study. This phase transition is associated with the emergence of a topological order, resulting from the pairing of vortices with opposite circulation. The BKT mechanism does not involve any spontaneous symmetry breaking or emergence of a spatially uniform order parameter. The low temperature phase is associated with a quasi-long-range order, at finite temperature, with the correlation of the order parameter decaying algebraically in space. Above the critical temperature the correlation decays exponentially.

For strong planar anisotropy in Hamiltonian (2.7) we have the so called large D phase. This phase consists of a unique ground state with total magnetization $S_{total}^z = 0$ separated by a gap from the first excited states which lie in the sectors

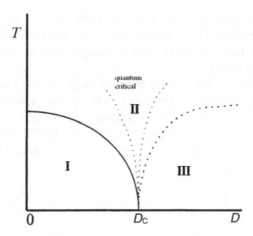

Figure 2.1. Phase diagram of the model described by the Hamiltonian (2.7). In regions I and III the physics can be described in terms of quasi-particle excitation. The excitations in region II do not have a particle-like character.

$S_{\text{total}}^z = \pm 1$. The elementary excitations are *excitons*, with $S = 1$ and an infinite lifetime at low energies. For small D the Hamiltonian (2.7) is in a gapless phase well described by the spin wave formalism. Increasing the anisotropy parameter D reduces the transition temperature and at D_C the critical temperature vanishes. Thus, at D_C the system undergoes a QPT, at $T = 0$, from a gapless to a gapped phase.

The phase diagram for the Hamiltonian (2.7) is expected to be as shown qualitatively in figure 2.1. The solid line in region I represents the line of critical points, determined by the BKT transition that terminates at the critical point D_C. Below this line, the inverse correlation length, ξ^{-1}, vanishes. The dashed line represents a crossover from the quantum critical region (where $\xi^{-1} \propto T$), to a region where quantum fluctuations dominate ($\xi^{-1} \gg T$).

At small D ($D < D_C$) and across the BKT transition, we can use a self-consistent harmonic approximation to describe the model. The large D phase can be studied using the $SU(3)$ Schwinger boson technique in a mean field approximation. For instance, we have found that the temperature dependence of the gap in this region is given by [5]

$$\Delta = \Delta_0 + c_1 T^{1/2} \exp(-c_2/T), \qquad (2.8)$$

where c_1 and c_2 are constants which depend on D, and Δ_0 is the gap at zero temperature. This same dependence was obtained using scaling arguments [3].

In the intermediate quantum critical regime neither description is adequate. A lack of well-defined weakly coupled quasi-particles in this region makes a description of quantum criticality difficult to study using traditional methods. The behavior at $T = 0$ can be understood using several methods such as the renormalization group technique, but the study of the dynamics at $T > 0$ in this region is quite complicated.

The renormalization group has also been applied but the results are not qualitatively accurate [3]. Outside of AdS/CFT there are no models of strongly coupled quantum criticality in $(2+1)$ dimensions in which analytic results for processes such as transport can be obtained [7].

In the low temperature phase of the region $D < D_C$ we have a power-law decay of the correlations but no broken symmetry. This phase cannot be described by a simple order parameter in the Landau–Ginzburg (LG) theory of phase transitions. We can, however, use the LG formalism to study the QPT if for $D < D_C$ we restrict ourselves only to the $T = 0$ limit. The low energy dynamics of the model is then described by the action [3, 4]:

$$S = \int d^d r d\tau \frac{1}{2}\left[c^2 (\nabla \phi)^2 + (\partial_\tau \phi)^2 + \delta \phi^2 + \frac{V}{2}\phi^4 \right], \qquad (2.9)$$

where here ϕ is a field variable and τ is an imaginary time ($\tau = it$). The transition is now tuned by varying $\delta \propto (D - D_C)$. This action, when written in terms of t, has an emergent symmetry not shared by the original theory in the lattice. It is invariant under Lorentz transformation, with the velocity c playing the role of the velocity of light. As is well known from the theory of phase transitions, at the critical point the action (2.9) is scale invariant.

This same action describes several other QPTs belonging to the same universality class. In dimensions $d < 4$, the theory is renormalizable, which implies that in the high energy limit it becomes free, while in the low energy limit the theory becomes strongly coupled, and the standard technique of perturbation in V is not valid.

2.3 Other models

Another simple model of interest is the system of spinless bosons hopping on a lattice of sites i with short range repulsive interactions described by the Hubbard Hamiltonian

$$H = -\omega \sum_{\langle ij \rangle} \left(b_i^+ b_j + b_i^+ b_j \right) + \frac{U}{2}\sum_i n_i (n_i - 1) \qquad (2.10)$$

where b_i is a boson annihilation operator, $n_i = b_i^+ b_i$, ω is the hopping matrix element between nearest-neighbor sites, and U is the on-site repulsive energy between a pair of bosons. Sachdev [2] has demonstrated that the superfluid–insulator QPT of the above Hamiltonian in two spatial dimensions is described by conformal field theory. Quantum critical points for metals in three dimensions are usually simpler, and the traditional perturbative theory appears to work.

Other examples of QPTs can be found in [7] and [3]. I will mention just a few here. The Néel–VBS (valence bond solid) transition in antiferromagnets, super-conducting–insulator transitions in thin metallic films, QPTs between strong Fermi liquids as found in heavy fermion intermetallics and high T_C superconductors, and a change from electron to hole character as a function of bias voltage in graphene.

At the critical point in the quantum rotor model (this model can be related to the XY model of section 2.2) that is used to describe the insulator–superconductor

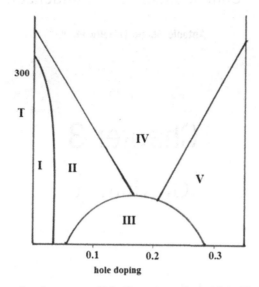

Figure 2.2. Phase diagram for the cuprate high T_C superconductor. I antiferromagnet, II pseudogap, III superconductor, IV strange metal, V Fermi liquid.

transition, the resistivity is linear in temperature. This property suggests that resistivity with this behavior can be associated with a quantum critical point, as in the case of high T_C superconductors. However, the role of QPTs in this system is up to now not well understood.

For the benefit of the reader figure 2.2 illustrates a schematic phase diagram for the cuprate high T_C superconductor, as a function of temperature and hole doping. Superconductivity and strange metals will be discussed, as mentioned earlier, in chapter 7. The pseudo gap is a state where, for instance, the Fermi surface is gapped only at certain points, while other parts retain their conducting properties. The meaning of regions I and V need not be elaborated on here.

References

[1] Sachdev S 2013 Strange and stringy *Sci. Am.* **308** 44
[2] Sachdev S 2012 What can gauge-gravity duality teach us about condensed matter physics *Annu. Rev. Cond. Matt. Phys.* **3** 9
[3] Sachdev S 1999 *Quantum Phase Transitions* (Cambridge: Cambridge University Press)
[4] Phillips P 2003 *Advanced Solid State Physics* (Boulder, CO: Westview Press)
[5] Pires A S T, Lima L S and Gouvea M E 2008 The phase diagram and critical properties of the two-dimensional anisotropic XY model *J. Phys.: Condens. Matter* **20** 015208
[6] Pires A S T and Gouvea M E 2009 Quantum phase transitions in the two-dimensional XY model with single-ion anisotropy *J. Phys. A: Math. Gen.* **388** 21
[7] Hartnoll S A 2009 Lectures on holographic methods for condensed matter physics *Class. Quantum Grav.* **26** 224002

Chapter 3

Gravitation

3.1 Basics of general relativity

In the general theory of relativity, the motion of matter is governed by the geometry of the space time, and this geometry, in turn, is determined by matter. In the equation proposed by Einstein, matter is described by the energy-momentum tensor, while curvature is represented by the Einstein tensor. In order to allow a stationary solution to his equations, Einstein added a term containing a constant Λ, known as the cosmological constant. When observations showed that the Universe was expanding, the idea of a static universe, and then the cosmological constant, was abandoned. Today it appears that the Universe is expanding at an increasing rate, and the cosmological constant has returned. Current observations of an accelerating universe are consistent with a de Sitter (dS) universe, differing from the AdS by the sign of the cosmological constant. We will consider mathematical universes, not necessarily real physical universes.

The reader does not need profound skills in general relativity in order to understand the main ideas presented here, unless calculations are required. In this section I will present a general overview of general relativity for the reader unfamiliar with the subject. (The reader familiar with the subject can skip this section.) An excellent book about general relativity is written by Carrol [1]. I will start with some definitions.

A contravariant vector V^μ is represented by upper indices and transforms under a coordinate transformation $x^\mu \rightarrow x'^\mu$ as

$$V'^\mu = V^\lambda \frac{\partial x'^\mu}{\partial x^\lambda}, \tag{3.1}$$

where I have used the Einstein convention: we sum over equal upper and lower indices.

A covariant vector V_μ is represented by lower indices and transforms as

$$V'_\mu = V_\lambda \frac{\partial x^\mu}{\partial x'^\lambda}. \tag{3.2}$$

A covariant or contravariant or mixed tensor has several indices which transform according to (3.1) or (3.2). An important object in non-Euclidean geometry is the line element given by

$$ds^2 = g_{\mu\nu} dx^\mu dx^\nu, \tag{3.3}$$

where $g_{\mu\nu}$ is called the metric tensor. It is common to use the term *metric* to describe ds^2. We define the inverse metric $g_{\mu\nu}$, via $g^{\mu\nu} g_{\nu\sigma} = \delta^\mu_\sigma$. The metric can be used to lower and raise indices: $x^\mu = g^{\mu\nu} x_\nu$, $x_\mu = g_{\mu\nu} x^\nu$. A contravariant (covariant) vector is transformed into a covariant (contravariant) one using the metric: $g_{\mu\nu} V^\nu = V_\mu$, $g^{\mu\nu} V_\nu = V^\mu$.

The metric determines the geometry of a space and as pointed out by Carrol [1] it has various uses: (i) it supplies a notion of 'past' and 'future'; (ii) it allows the computation of path length and proper times; (iii) it determines the 'shortest distance' between two points, and therefore the motion of test particles; (iv) it replaces the Newtonian gravitational field ϕ.

The metric tensor contains all the information we need to describe the curvature of the space time.

In non-Euclidean geometry a vector changes its components under parallel transport. If we go from x_μ to $x_\mu + \varepsilon$, we have

$$V_\nu(x) \rightarrow V_\nu(x + \varepsilon) - \varepsilon \Gamma^\rho_{\mu\nu} V_\rho(x), \tag{3.4}$$

where $\Gamma^\rho_{\mu\nu}$ are called the Christofell symbols or connection coefficients. The derivative of a vector does not form a tensor. We define the covariant derivative (which is a covariant tensor of rank 2) by

$$\nabla_\mu V_\lambda \equiv \partial_\mu V_\lambda - \Gamma^\rho_{\mu\lambda} V_\rho. \tag{3.5}$$

We can show that

$$\Gamma^\lambda_{\mu\nu} = \frac{1}{2} g^{\lambda\sigma} \left(\partial_\mu g_{\nu\sigma} + \partial_\nu g_{\sigma\mu} - \partial_\sigma g_{\mu\nu} \right). \tag{3.6}$$

The covariant derivative of a scalar is just the partial derivative. The next object of interest is the Riemann tensor

$$R^\rho_{\sigma\mu\nu} = \partial_\mu \Gamma^\rho_{\nu\sigma} - \partial_\nu \Gamma^\rho_{\mu\sigma} + \Gamma^\rho_{\mu\lambda} \Gamma^\lambda_{\nu\sigma} - \Gamma^\rho_{\nu\lambda} \Gamma^\lambda_{\mu\sigma}, \tag{3.7}$$

and

$$R_{\rho\sigma\mu\nu} = g_{\rho\lambda} R^\lambda_{\sigma\mu\nu}. \tag{3.8}$$

The Riemann tensor can be used to derive two quantities. The first is the Ricci tensor, which is calculated by contraction on the first and third indices:

$$R_{\mu\nu} = R^{\lambda}_{\mu\lambda\nu}. \tag{3.9}$$

Using contraction on the Ricci tensor, we obtain the Ricci scalar

$$R = R^{\mu}_{\mu} = g^{\mu\nu} R_{\mu\nu}. \tag{3.10}$$

Einstein's equations for the gravity are then written as

$$R_{\mu\nu} - \frac{1}{2} R g_{\mu\nu} + \Lambda g_{\mu\nu} = 8\pi G_{\mathrm{N}} T_{\mu\nu}, \tag{3.11}$$

where Λ is the cosmological constant, G_{N} Newton's cosmological constant, and $T_{\mu\nu}$ the *stress-energy* or *energy-momentum* tensor. This tensor acts as the source of the gravitational field. The T^{00} component represents energy density. The T^{i0} component is the momentum density in the i direction. The component T^{0i} (which is equal to T^{i0}) represents the energy flux across the surface x^i. And finally T^{ij} is the *stress*.

Calculation of the Christoffel symbols, the Riemann tensor and other objects, starting from the metric, can be performed using programs such as *Maple* or *Mathematica*. (In *Maple*, the *Tensor package* contains routines that deal with tensors in general relativity both on a natural basis and in a moving frame.)

In principle, from Einstein's equation (3.11) it would be possible to derive the metric for a given distribution of matter and energy. However this is quite a difficult task. What is usually done is to start from an ansatz metric, exploring the symmetry of the model with some unknown coefficients. $R_{\mu\nu}$ is then derived and using equation (3.11) the unknown coefficients are determined. So far this procedure has been used only for very special, although physically relevant, symmetric conditions (see appendix A for an example).

In empty space (i.e. $T_{\mu\nu} = 0$) equation (3.11) becomes

$$R_{\mu\nu} - \frac{1}{2} R g_{\mu\nu} + \Lambda g_{\mu\nu} = 0. \tag{3.12}$$

Multiplying (3.12) by $g^{\mu\nu}$, using equation (3.10) and $g^{\mu\nu} g_{\mu\nu} = n$, where n is the dimension of the space time, we find

$$R = \frac{2n}{n-2} \Lambda. \tag{3.13}$$

Taking (3.12) into (3.11) we obtain

$$R_{\mu\nu} = \frac{2\Lambda}{n-2} g_{\mu\nu}. \tag{3.14}$$

Einstein's equations 'in vacuum' can be obtained by varying the following action (called the Hilbert action) with respect to the metric

$$S = \int \mathrm{d}^4x \sqrt{g} \left[\frac{1}{16\pi G_{\mathrm{N}}} (R - 2\Lambda) \right]. \tag{3.15}$$

The term $\sqrt{g}$, where g is the modulus of the determinant of the metric, is necessary, since only $\mathrm{d}^4x\sqrt{g}$ is invariant under a general coordinate transformation. For a demonstration of the derivation see [1]. The energy momentum tensor can be obtained from the total Lagrangian, including a Lagrangian for the matter, by variation with respect to the inverse metric (see appendix B):

$$T_{\mu\nu} = \frac{\delta L_{\text{total}}}{\delta g^{\mu\nu}}. \tag{3.16}$$

Hilbert action is appropriate only for a closed space time. When the underlying space time has a boundary (such as the AdS space) one needs to add a term called the Gibbons–Hawking boundary term [2], so that the variational principle is well defined. This term affects which boundary conditions we impose on the metric, in the same way that the $\int \phi \vec{h}\cdot\vec{\partial}\phi$ term in the scalar case changes the scalar boundary condition from Neumann to Dirichlet. The term is written as

$$S_{\text{GH}} = -\frac{1}{8\pi G_{\text{N}}}\int_{\partial M}\mathrm{d}^3x\sqrt{\gamma}K, \tag{3.17}$$

where γ is the induced metric on the boundary surface, K is the trace of the extrinsic curvature of the boundary

$$K = \gamma^{\mu\nu}\nabla_\mu n_\nu = \frac{n^r}{2}\gamma^{\mu\nu}\partial_r\gamma_{\mu\nu}, \tag{3.18}$$

and n^r is the outward-pointing unit normal to the boundary.

At any point p of the space time there exists a coordinate system in which the metric takes the canonical form $g_{\mu\nu} = \mathrm{dig}\,(-1, +1, +1, +1)$ in $d = 4$ for instance. Such coordinates are known as locally inertial coordinates, and the associated vectors constitute a local Lorentz frame. We can simultaneously construct sets of basis vectors at every point in the space time such that the metric takes the canonical form. The problem is that in general there will not be a coordinate system from which this basis can be derived. We can use sets of basis vectors that are not derived from any coordinate systems, and are more convenient in gauge theories of physics. Let us imagine that at each point in the space time we introduce a set of basis vectors $\hat{e}_{(a)}$. These basis vectors can be chosen to be 'orthonormal'. The set of vectors comprising an orthonormal basis is sometimes known as vielbein (from the German for 'many legs') or tetrad. On a non-coordinate basis we replace the ordinary connection coefficients $\Gamma^\lambda_{\mu\nu}$ by the so called spin connection, denoted by $\omega^a_{\mu b}$. It is usual to use Latin letters to remind us that they are not related to any coordinate system. The spin connection is used in the Dirac equation in curved space time in section 7.2, since fermions are described by a spinor and not a scalar quantity.

To conclude this section let us look at the action of free electrodynamics in non-Euclidean geometry. In three spatial dimensions this is given by

$$S = \int \mathrm{d}^4x\sqrt{g}L, \tag{3.19}$$

where

$$L = F_{\mu\nu}F_{\alpha\beta}g^{\mu\alpha}g^{\nu\beta} = F_{\mu\nu}F^{\mu\nu}. \tag{3.20}$$

The electromagnetic field tensor is given by

$$F_{\mu\nu} = D_\mu A_\nu - D_{\nu\mu} = \partial_\mu A_\nu - \partial_\nu A_\mu - \left(\Gamma^\sigma_{\mu\nu} - \Gamma^\sigma_{\nu\mu}\right)A_\sigma = \partial_\mu A_\nu - \partial_\nu A_\mu, \tag{3.21}$$

where I have used the symmetry relation: $\Gamma^\sigma_{\mu\nu} = \Gamma^\sigma_{\nu\mu}$. Therefore we do not need to worry about the covariant derivative. The Euler–Lagrange equation is thus given by

$$\partial_\rho \frac{\partial L}{\partial(\partial_\rho A_\sigma)} - \frac{\partial L}{\partial A_\sigma} = 0. \tag{3.22}$$

For the Lagrangian equation (3.20) this becomes

$$\frac{\partial L}{\partial A_\sigma} = 0. \tag{3.23}$$

Equation (3.20) gives

$$\partial_\rho \frac{\partial L}{\partial(\partial_\rho A_\sigma)} = \partial_\rho \frac{\partial}{\partial(\partial_\rho A_\sigma)}\left(\sqrt{g}\, F_{\mu\nu}F_{\alpha\beta}\, g^{\mu\alpha}g^{\nu\beta}\right)$$

$$= \partial_\rho\left[\sqrt{g}\, g^{\mu\alpha}g^{\nu\beta}\frac{\partial}{\partial(\partial_\rho A_\sigma)}\left(F_{\mu\nu}F_{\beta\beta}\right)\right]. \tag{3.24}$$

Performing the calculation explicitly we find

$$\frac{\partial L}{\partial(\partial_\rho A_\sigma)} = \sqrt{g}\, g^{\mu\alpha}g^{\nu\beta}\frac{\partial}{\partial(\partial_\rho A_\sigma)}\left[2\left(\partial_\mu A_\nu - \partial_\nu A_\mu\right)\partial_\alpha A_\beta\right]$$

$$= 2\sqrt{g}\, g^{\mu\alpha}g^{\nu\beta}\left[\delta^\rho_\mu \delta^\sigma_\nu \partial_\alpha A_\beta - \delta^\rho_\nu \delta^\sigma_\mu \partial_\alpha A_\beta + \left(\partial_\mu A_\nu - \partial_\nu A_\mu\right)\delta^\rho_\alpha \delta^\sigma_\beta\right]$$

$$= \sqrt{g}\left[2g^{\rho\alpha}g^{\sigma\beta}\partial_\alpha A_\beta - 2g^{\sigma\alpha}g^{\rho\beta}\partial_\alpha A_\beta + g^{\alpha\rho}g^{\beta\sigma}\left(\partial_\alpha A_\beta - \partial_\beta A_\alpha\right)\right]$$

$$= 4\sqrt{g}\, g^{\mu\rho}g^{\nu\sigma}\left(\partial_\mu A_\nu - \partial_\nu A_\mu\right). \tag{3.25}$$

Inserting this into equation (3.26) we get the Maxwell equation

$$g^{\mu\rho}g^{\nu\sigma}\partial_\rho F_{\mu\nu} = -\frac{1}{\sqrt{g}}F_{\mu\nu}\partial_\rho\left(\sqrt{g}\, g^{\mu\rho}g^{\nu\sigma}\right), \tag{3.26}$$

or

$$\frac{1}{\sqrt{g}}\partial_\rho\left(\sqrt{g}\, F^{\alpha\rho}\right) = 0. \tag{3.27}$$

3.2 Anti-de Sitter space

Anti-de Sitter (AdS) space belongs to a wide class of homogeneous spaces that can be defined as quadric surfaces in flat vector spaces. The standard example is the d-dimensional sphere S^d defined by

$$X_1^2 + \cdots + X_{d+1}^2 = L^2, \tag{3.28}$$

embedded in a Euclidean $(d+1)$-dimensional space. The d-dimensional de Sitter space is defined by

$$-X_0^2 + X_{d+1}^2 + \sum_{i=1}^{d-1} X_i^2 = L^2, \tag{3.29}$$

embedded in a flat $(d+1)$-dimensional space with metric

$$ds^2 = -dX_0^2 + dX_{d+1}^2 + \sum_{i}^{d-1} dX_i^2. \tag{3.30}$$

It has constant positive curvature. The d-dimensional AdS space can be defined as the quadric

$$X_0^2 + X_{d+1}^2 - \sum_{i=1}^{d-1} X_i^2 = L^2, \tag{3.31}$$

embedded in a flat $(d+1)$-dimensional space with the metric

$$ds^2 = -dX_0^2 - dX_{d+1}^2 + \sum_{i=1}^{d-1} dX_i^2. \tag{3.32}$$

AdS space is a maximally symmetric solution of the Einstein equation in the absence of matter and energy with a negative cosmological constant. A maximally symmetric Lorentz manifold corresponds to a space time in which time and space in all directions are mathematically equivalent; it has constant negative curvature. Negative curvature means curved hyperbolically, like a saddle surface or the one shown in figure 3.1. The metric can be written in different forms, corresponding to different coordinate systems.

In Poincaré coordinates $(u, t, \vec{x})(0 < u, \vec{x} \epsilon R^{d-2})$ defined by [3]:

$$X_0 = \frac{1}{2u}\left[1 + u^2(L^2 + \vec{x}^2 - t^2)\right]$$

$$X_d = Lut,$$

$$X^i = Lux^i \qquad (i = 1, \ldots, d-2), \tag{3.33}$$

$$X^{d-1} = \frac{1}{2u}\left[1 - u^2(L^2 - \vec{x}^2 + t^2)\right],$$

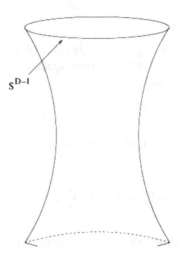

Figure 3.1. Image of a two-dimensional anti-de Sitter space embedded in a flat three-dimensional space.

the AdS metric becomes

$$ds^2 = L^2 \left[\frac{du^2}{u^2} + u^2 \left(-dt^2 + d\vec{x}^2 \right) \right]. \tag{3.34}$$

The Poincaré coordinates cover only part of the AdS space. Another equivalent form of the metric is obtained from (3.34) by setting $r = 1/u$, leading to

$$ds^2 = \frac{L^2}{r^2} \left[-dt^2 + dr^2 + d\vec{x}^2 \right]. \tag{3.35}$$

For simplicity let us consider the metric

$$ds^2 = \frac{L^2}{r^2} \left(dr^2 + dx^2 \right). \tag{3.36}$$

This is the Poincaré half-plane, which is the region $r > 0$ of a two-dimensional region with coordinates (x, r). It is easy to verify that the non-vanishing Christoffel symbols for (3.36) are $\Gamma^x_{xr} = \Gamma^x_{rx} = -r^{-1}$, $\Gamma^r_{xx} = r^{-1}$, $\Gamma^r_{rr} = -r^{-1}$. A representative component of the Riemann tensor is $R^x_{rxr} = -r^{-2}$. All other components are either zero or related to this by symmetry. The Ricci tensor is $R_{xx} = -r^{-2}$, $R_{xr} = 0$, $R_{rr} = -r^{-2}$, and the curvature scalar is $R = -2/L^2$. The length of a line segment stretching vertically, with $x = \text{constant}$, from r_1 to r_2 is given by

$$\Delta s = \int_{r_1}^{r_2} \sqrt{g_{\mu\nu} \frac{dx^\mu}{dr} \frac{dx^\nu}{dr}} \, dr = L \int_{r_1}^{r_2} \frac{dr}{r} = L \ln \left(\frac{r_2}{r_1} \right). \tag{3.37}$$

We can see that the path length becomes infinite when $r \to 0$, meaning that the 'boundary' is located at $r = 0$. Such a boundary is called the *conformal boundary*. Keeping x fixed and moving in the r direction from a finite value of r to $r = 0$ is actually an infinite distance. However the light (in a space time AdS) travels along null geodesics and reaches the boundary in finite time. The same procedure can be extended to higher dimensions. For the case of the space time of general relativity the conformal boundary has an induced geometry of Minkowski space.

The idea of AdS/CFT correspondence is that the conformal boundary of AdS can be taken as a space time for quantum conformal field theory that is equivalent to classical gravitational theory on the bulk of the AdS, as will be shown in more detail later.

References

[1] Carrol S M 2004 *Spacetime and Geometry* (Reading, MA: Addison-Wesley)
[2] Gibbons G W and Hawking S W 1977 Action integrals and partition functions in quantum gravity *Phys. Rev.* D **15** 2752
[3] Harony O, Gubser S S, Maldacena J, Oooguri H and Oz Y 2000 Large N field theories, string theory and gravity *Phys. Rep.* **323** 183

Chapter 4

AdS/CFT correspondence

4.1 Formalism

At the fixed point, the theory describing a QPT is invariant under rotation, translation, time translation and dilatation. The theory is also scale invariant. This means that the transformation $x^\mu \to \lambda x^\mu$ ($\mu = 0, 1, 2, d - 1$) is symmetric. A conformal transformation is a local change of scale: $d\tilde{s}^2 = \Omega(x)\,ds^2$, where $\Omega(x)$ is an arbitrary function of the coordinates. Scale transformations are a particular case of conformal transformation with the constant $\Omega = \lambda$. (The more general case, $t \to \lambda^z t$, where z is the dynamical critical exponent, will not be considered here. This is referred to as Lifshitz scaling in the literature.)

We now look for a metric, one dimension higher than the field theory in which these symmetries are realized. Note that to have translational invariance in the t and x^i directions, we have to assume that all metric components depend on the extra dimension only. One metric with these properties is given by:

$$ds^2 = \frac{L^2}{r^2}(-dt^2 + dr^2 + dx^i dx^i). \tag{4.1}$$

We see that this metric (4.1) remains unchanged if we multiply r, t and x^i by λ. The coordinates $\{t, x^i\}$ parameterize the space on which the field theory lies, while r is the extra coordinate running from $r = 0$ (the 'boundary') to $r = \infty$ (the 'horizon'). This is the metric for the AdS space time as was discussed in section 3.2. (For a general z, the term in dt^2 in the metric is written as $-dt^2/r^{2z}$.) The indices $\{\mu, \nu\}$ will run over d space time dimensions for the QFT, indices $\{i, j\}$ over the spatial coordinates x^i, and $\{A, B\}$ over the full $(d + 1)$-dimensional bulk with the parameter L setting the radius of curvature for the AdS space time. As was mentioned in section 1.2, weak curvature (large L) corresponds to strong coupling in field theory. The AdS manifold is a collection

4-1

of copies of d-dimensional Minkowski space of varying size (constant r slices are just copies of Minkowski space). Of course this metric can be written in different forms using different coordinates and in the above equation we have used Poincaré coordinates. The effect of bulk quantum corrections is to renormalize the value of L.

Taking the metric (4.1) into the Einstein equation of motion (3.12) we obtain $\Lambda = -d(d+1)/2L^2$. Inserting this relation into Einstein's equation for an empty space and a negative cosmological constant, equation (3.14), and remembering that $n = d + 1$, we arrive at

$$R_{ab} = -\frac{d}{L^2}g_{ab}. \tag{4.2}$$

The metric (4.1) also enjoys Lorentz boost symmetry and special conformal symmetries. It is believed that the classical dynamics about this background metric describe the physics of the strongly coupled theory. In other texts the inverse of the radial coordinate is used and the coordinate used here is denoted by z. Following [1] z is kept free to denote the dynamical critical exponent.

As pointed out by McGreevy [2] the extra dimension of space time in the metric (4.1) has a clear physical meaning (figure 4.1). In field theory, the renormalization group (RG) equation for the behavior of the coupling constants as a function of the RG scale u is [3]

$$u\frac{\partial g}{\partial u} = \beta[g(u)], \tag{4.3}$$

where the symbol β is used for historical reasons. This suggests that we can take the extra dimension as an energy scale. At the critical point we have $\beta = 0$ and the system

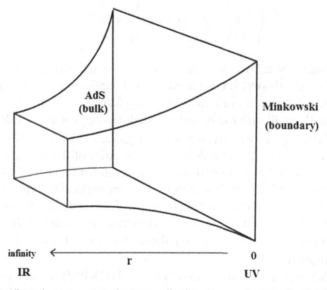

Figure 4.1. The dimension r represents the renormalization group scale. UV: ultraviolet, IR: infrared.

is self-similar. The field theory becomes invariant under conformal transformation. Dimensional analysis says that u will scale under the scale transformation as $u \rightarrow u/\lambda$. We can perform a change of coordinate $r \equiv L^2/u$ and arrive at the metric in (4.1). Gauge gravity duality has transformed the renormalization invariance of quantum critical theory into a geometrical invariance of the metric. Solving the gravitational equations of motion is therefore in duality with following the RG flow in energy scales. Thus a field theory process with a characteristic energy E is associated with a bulk process at $\sim 1/E$.

Maldacena [4] conjectured the AdS/CFT duality, but his conjecture did not specify the precise way in which these two theories should be mapped onto each other. Later a proposal was made by Gubser, Klebanov and Polyakov [5] and by Witten [6]. The formulation is as follows.

As is well known, in condensed matter an important ingredient is the partition function Z which is related to the free energy F by $Z = \exp[-\beta F]$. It will be shown in section 5.1 that if we perturb the system with an external field (called the source) J coupled to a local operator O (called the response) the partition function is given by

$$Z = tr e^{-\beta\left[H - \int d\vec{x} O J\right]} \qquad (4.4)$$

where H is the unperturbed Hamiltonian. From Z we can obtain all the information about the equilibrium properties of the system. Let $\phi(x, r)$ be a dynamical gauge field in the bulk whose value on the boundary is equal to the source field of the boundary field theory, that is $\phi(x, r)|_{r=0} \rightarrow J(x) = \phi_0(x)$. The map between AdS and CFT quantities is then given by

$$\left\langle \exp\left(\int_{\text{boundary}} d^d x \phi_0 O \right) \right\rangle = e^{-S[\phi]}, \qquad (4.5)$$

where the left-hand side (which is the generating functional Z for the correlation functions in the field theory) is calculated in the boundary, and the exponent on the right-hand side is the action evaluated using the classical solution to the equation of motion for ϕ in the bulk with the boundary condition $\phi|_{r=0} = \phi_0$. A calculation using the AdS/CFT correspondence involves solving classical equations of motion for the fields $\phi(x, r)$ using a metric that reflects the symmetries of the boundary theory.

There is a field ϕ for every operator O in this dual field theory. Two examples of the correspondence (that are determined by symmetry) between bulk fields and boundary operators are: the stress-energy tensor $T_{\mu\nu}$ is the response of a local QFT to a local change in the metric $(g_{\mu\nu} \leftrightarrow T_{\mu\nu})$; electromagnetic (gauge) fields in the bulk correspond to currents in the boundary theory $(A_\mu^a \leftrightarrow J_a^\omega)$. Later I will show how given ϕ, O can be found.

Gubser, Klebanov, Polyakov [5] and Witten [6] (GKPW) found a dictionary that relates the physics on both sides of the duality. The rules are simple, but the dictionary is considered mysterious even by the professionals. There is evidence that the rules are true, but as yet there is no mathematical proof. The dictionary is:

bulk	↔	boundary
fields		local operators
spin		spin
mass		scaling dimension Δ
metric $g_{\mu\nu}$		energy momentum tensor $T^{\mu\nu}$
gauge field A_μ		conserved current J^μ
black hole		deconfined phase at temperature T
charged black hole		chemical potential
gauge symmetry		global symmetry

The dictionary is simple; the difficulty is how to figure out the holographic dualities. Note that there is a requirement for the bulk to have AdS geometry near the boundary. This requirement can be violated in the interior of the bulk, as we will see later.

4.2 Scalar fields

As has already been mentioned, we solve the classical equation of motion for a field ϕ in the bulk in duality with a QFT in the boundary. Starting with classical gravity in the bulk, the Lagrangian of the vast majority of dual field theories in the boundary is not known explicitly. Rather than a Lagrangian, dual field theories are characterized by their spectrum of operators and the correlation functions of these operators. In this section we will consider a simple example of a scalar field. We will present all the details of the calculations, since models studied later will follow the same procedure.

The action for a free scalar field in Minkowski space is given by:

$$S = -\frac{1}{2}\int \mathrm{d}x^4\left(\eta^{\mu\nu}\partial_\mu\phi\partial_\nu\phi + m^2\phi^2\right), \tag{4.6}$$

where $\eta^{\mu\nu}$ is the diagonal matrix

$$\eta^{\mu\nu} = \begin{pmatrix} -1 & 0 & 0 & 0 \\ 0 & 1 & 0 & 0 \\ 0 & 0 & 1 & 0 \\ 0 & 0 & 0 & 1 \end{pmatrix}. \tag{4.7}$$

Moving to curved space time we replace:

$$\eta^{\mu\nu} \to g^{\mu\nu} \qquad \mathrm{d}x^4 \to \sqrt{g}\,\mathrm{d}x^4 \qquad \partial \to \nabla, \tag{4.8}$$

where the meanings of g and ∇ were introduced in chapter 3. However, for a scalar field the covariant derivative D is just the normal one. So the action for a scalar field in the bulk is given by (the field ϕ changes the metric, it back reacts, but this effect will be neglected, see appendix B):

$$S = -\frac{1}{2}\int \mathrm{d}^{d+1}x\sqrt{g}\left(g^{AB}\partial_A\phi\partial_B\phi + m^2\phi^2\right), \tag{4.9}$$

where as before $\sqrt{g} = \sqrt{|\det g|} = (L/r)^{d+1}$. The field ϕ is governed by classical field theory in the curved AdS space time, with the metric given by equation (4.1). It is assumed that we add to (4.9) the Hilbert action that gives the metric. If we are not considering the effect of the field in the metric, it is irrelevant to add the Hilbert term. We just use the metric given by this term. However if we are taking into account the effect of back reaction we have to use the full Lagrangian. Note that in our case of a diagonal metric, we have $g^{AA} = 1/g_{AA}$. We remark that masses in the bulk do not correspond to masses in the dual field theory. Note also that the action (4.9) is the dual of the one for a strongly coupled system at the boundary and not the dual of the action (4.6), which is for free particles and, therefore, can be treated easily.

The calculation below is representative of the technical work associated with applying AdS/CFT correspondence to condensed matter physics. The field equation for ϕ is given by (see appendix B):

$$\frac{1}{\sqrt{g}}\partial_A\left(\sqrt{g}\,g^{AB}\partial_B\phi\right) - m^2\phi = 0, \tag{4.10}$$

which leads to the following equation of motion

$$\left[r^{d+1}\partial_r\left(\frac{1}{r^{d-1}}\partial_r\right) + r^2\left(-\partial_t^2 + \nabla^2\right) - m^2 L^2\right]\phi = 0. \tag{4.11}$$

We see that the interaction with the AdS metric leads to an interaction term in (4.11), which vanishes in Minkowski space. Integrating by parts we can write the action (4.9) as

$$S = -\frac{1}{2}\int d^{d+1}x\left[\partial_A\left(\sqrt{g}\,g^{AB}\phi\partial_B\phi\right) - \phi\partial_A\left(\sqrt{g}\,g^{AB}\partial_B\phi\right) + \sqrt{g}\,m^2\phi^2\right]. \tag{4.12}$$

Using Stoke's theorem we find

$$S = -\frac{1}{2}\int_{\text{boundary}} d^d x\sqrt{g}\,g^{rr}\phi\partial_r\phi + \frac{1}{2}\int dx^{d+1}\phi\left[\partial_A\sqrt{g}\,g^{AB}\partial_B - m^2\right]\phi. \tag{4.13}$$

If ϕ solves the equation of motion, the first integral vanishes and the action on the shell is given just by the boundary term. Translation invariance allows us to perform a Fourier transform in directions other than r. Writing $\phi(\vec{x}, r, t) = e^{-(\omega t - \vec{k}\cdot\vec{r})}f(r)$ we find

$$\left[r^{d+1}\partial_r(r^{-d+1}\partial_r) - r^2(\omega^2 - k^2) - m^2 L^2\right]f(r) = 0. \tag{4.14}$$

We first consider asymptotic solutions for this equation near the boundary $r = 0$, writing $f(r) \propto r^\Delta$. Taking this expression into equation (4.11) we obtain

$$\Delta(\Delta - d)r^\Delta + r^{2+\Delta}(\omega^2 - k^2) - m^2 L^2 r^\Delta = 0 \tag{4.15}$$

which for $r \to 0$ leads to

$$\Delta(\Delta - d) = m^2 L^2, \tag{4.16}$$

with solutions

$$\Delta_{\pm} = \frac{d}{2} \pm \sqrt{\left(\frac{2}{2}\right)^2 + m^2 L^2}. \tag{4.17}$$

We see that $\Delta_+ \equiv \Delta$ is always positive, therefore r^Δ decays near the boundary and the leading behavior near the boundary of any general solution is $\phi \propto r^\Delta$. Real exponents imply the so called Breitenlohner–Freedman condition: $m^2 L^2 \geqslant -d^2/4$. This is a property of the AdS space. As long as this condition is satisfied the space is stable in the presence of the massive field ϕ. If $m^2 L^2 < -d^2/4$, one has a linear instability (corresponding to the presence of 'tachyons' in string theory). Noting that equation (4.11) can be written as

$$\left[r^2 \frac{d^2}{dr^2} + (1 - d)r\frac{d}{dr} - r^2 k^2 - m^2 L^2 \right] f_k(r) = 0, \tag{4.18}$$

and using the Bessel equation

$$x^2 \frac{d^2 J_n(x)}{dx^2} + x\frac{d J_n(x)}{dx} + (x^2 - n) J_n(x) = 0, \tag{4.19}$$

we see that writing $k^2 = \omega^2 - \mathrm{k}^2$, the solution of equation (4.18) for $k^2 > 0$ is given by

$$f_k(r) = A r^{d/2} K_\nu(kr) + B r^{d/2} I_\nu(kr), \tag{4.20}$$

where K_ν and I_ν are modified Bessel functions, and

$$\nu = \Delta - d/2 = \sqrt{(d/2)^2 + m^2 L^2}. \tag{4.21}$$

We choose K over I because it is well behaved near the horizon, $K(x) \approx e^{-x}$ at large x. As $r \to 0$, $K_\nu(r) \propto r^{-\nu}$. Therefore near the boundary, the bulk solution will behave like:

$$\phi(r) = \phi_{(0)} r^{d-\Delta} \qquad \text{as } r \to 0. \tag{4.22}$$

Let us analyze equation (4.22). We want information about the field theory at the boundary ($r = 0$), but ϕ vanishes there. There is no 'contact' at the boundary. The CFT information is encoded in the near boundary asymptotes (see discussion following equation (4.34)).

Choosing a cut off at $r = \varepsilon$ near $r = 0$ (to avoid divergences) we can normalize the solution (4.20) using the condition $f_k(r = \varepsilon) = 1$, and write

$$f_k(r) = \frac{r^{d/2} K_\nu(kr)}{\varepsilon^{d/2} K_\nu(k\varepsilon)}, \tag{4.23}$$

which is the 'bulk-to-boundary propagator'. The general solution to equation (4.11) can be written as

$$\phi(r, x) = \int \frac{d^d k}{(2\pi)^d} e^{ik \cdot x} f_k(r) \phi_0(k), \qquad (4.24)$$

where $\phi_0(k)$ is determined by the boundary condition

$$\phi(\varepsilon, x) = \int \frac{d^d k}{(2\pi)^d} e^{ik \cdot x} \phi_0(k). \qquad (4.25)$$

The field ϕ is a classical solution that is regular in the bulk and asymptotes to a given value ϕ_0 at the boundary. The above solutions are generic for classical free fields in the bulk. The action on the shell reduces to the surface terms

$$S = \frac{1}{2} \int \frac{d^d k}{(2\pi)^d} \phi_0(-k) \Im(k, r) \phi_0(k) \Big|_{r=\varepsilon}^{r=\infty}, \qquad (4.26)$$

where

$$\Im(k, r) = \sqrt{g} g^{rr} f_k(r) \partial_r f_k(r). \qquad (4.27)$$

Any n-point correlation function (also called the Green's function or propagator—see chapter 5) can be computed by varying the logarithm of the partition function including the source terms, and taking the limit of vanishing source strength. For instance

$$\langle O \rangle \equiv \frac{\delta \ln Z}{\delta J(x)} \Big|_{J=0}. \qquad (4.28)$$

The two-point correlation function evaluated using the equation

$$\langle\langle O(x) O(y) \rangle\rangle = \frac{\delta}{\delta J(x)} \frac{\delta}{\delta J(y)} \ln Z|_{J=0}, \qquad (4.29)$$

becomes

$$\langle\langle O(k_1) O(k_2) \rangle\rangle = -\frac{\delta}{\delta \phi_0(k_1)} \frac{\delta}{\delta \phi_0(k_2)} S = (2\pi)^d \delta^d (k_1 + k_2) \Im(k_1, \varepsilon). \qquad (4.30)$$

In QFT it is usual to call the propagator, given by (4.29), the correlation function and use the notation $\langle ... \rangle$. However, in condensed matter the correlation function, with the notation $\langle ... \rangle$, has another meaning, as will be explained in section 5.1. The calculation of equation (4.30) is presented in reference [7] and the final result is

$$\langle\langle OO \rangle\rangle_{k,\omega} \propto \left(k^2 + \omega^2 \right)^{\Delta_+ - d/2}. \qquad (4.31)$$

It is known that if for large x we have

$$\langle\langle O(x)O(0)\rangle\rangle = \frac{1}{|x|^{2h}}, \tag{4.32}$$

the parameter h is called the scaling dimension of the operator O. Fourier transforming (4.32) we find

$$\langle\langle OO\rangle\rangle \propto k^{2(h-d/2)}. \tag{4.33}$$

Comparing with equation (4.31) we see that Δ can be interpreted as the scaling dimension of the operator O in boundary theory. We say that an operator O has a scaling dimension Δ if it satisfies $O(x) \rightarrow \lambda^{\Delta}O(x)$ when we apply the transformation $t \rightarrow \lambda^{z}t$, $x \rightarrow \lambda x$. If the transformation makes the operator more important when we move to lower energy (low energy corresponds to large values of r, the IR region in figure 4.1) we call the operator relevant, if less important we call it irrelevant, and if the importance of the operator stays the same, then we have a marginal operator. From equation (4.16) we see that a simple bulk theory is the dual of a CFT with a hierarchy of operator dimensions.

If $\Delta \leqslant d$ (i.e. $d - \Delta \geqslant 0$) the operator O is relevant or marginal. Relevant operators do not destroy the asymptotically AdS region of the metric [8].

We are now ready to present a general rule proposed by GKPW [5, 6]: in using the AdS/CFT correspondence we consider on-shell action with the boundary value of the field to be equal to the source in the dual field theory: $\phi(r, x) \rightarrow J(x)$. When we substitute for ϕ a solution for the equation of motion, the 'bulk' term integrated over the whole of the AdS space, vanishes. This solution has a universal asymptotic behavior near the boundary $r = 0$ as

$$\phi(\omega, k, r) = A(\omega, k)r^{\Delta_-} + B(\omega, k)r^{\Delta_+} + \cdots \tag{4.34}$$

where $\Delta_{\pm} = d/2 \pm \nu$, and ν is given by equation (4.21). We can identify A as the source. The term Δ_- is in general negative and thus the boundary value for ϕ is not well defined. To avoid divergences one should compute at an infinitesimal distance $r = \varepsilon$ away from the boundary and then take the limit $\varepsilon \rightarrow 0$. The expectation value $\langle O \rangle$ for the field theory operator is given by

$$\langle O(\omega, k)\rangle = 2\nu B(\omega, k). \tag{4.35}$$

The two-point (normalized) propagator for conformal operators with scaling dimension Δ_+ is given by

$$\langle\langle O(-w, -k)O(w, k)\rangle\rangle = 2\nu\frac{B(\omega, k)}{A(\omega, k)}. \tag{4.36}$$

The boundary conditions fix B in terms of A and thereby determine the correlation function (4.36) up to an overall normalization.

Applying equation (4.35) to the result (4.20) we obtain

$$\langle\langle O(\omega, k)O(0)\rangle\rangle = 2\nu\frac{\Gamma(-\nu)}{\Gamma(\nu)}\left(\frac{ik}{2}\right)^{2\nu},\tag{4.37}$$

where Γ is the gamma function. Equation (4.37) agrees with the result found using a standard QFT procedure [9, 10].

As mentioned earlier, the holographic procedure does not give a Lagrangian for the boundary theory, it furnishes only $\langle O\rangle$ and the correlation functions.

4.3 Finite temperature

Finite temperature or the chemical potential breaks the dilatation symmetry of the space time. However, we expect that the space time should recover scaling invariance as we move towards the boundary, that is, the space time should be asymptotically AdS. If we relax the scaling symmetry but wish to preserve spatial rotations and space time translations, we start with the metric written as

$$ds^2 = \frac{L^2}{r^2}\left(-f(r)dt^2 + g(r)dr^2 + h(r)dx^idx^i\right).\tag{4.38}$$

If $f\neq h$ the metric breaks Lorentz invariance, as is the case for finite temperature or a finite chemical potential. We know that all QFT can be placed at temperatures other than zero. Therefore we do not need a new ingredient here and can use the same action as the one used in the $T=0$ case. Let us write (with $L=1$):

$$g_{tt} = -\frac{f(r)}{r^2} \equiv -\tilde{f}(r) \qquad g_{rr} = \frac{g(r)}{r^2} \equiv \tilde{g}(r) \qquad g_{xx} = g_{yy} = \frac{h(r)}{r^2} = \tilde{h}(r).\tag{4.39}$$

The metric is diagonal, and therefore the inverse metric is given by $g^{\mu\mu} = 1/g_{\mu\mu}$. Using equation (3.6) we have, for instance:

$$\Gamma^r_{rr} = \frac{1}{2}g^{r\rho}\left(\partial_r g_{r\rho} + \partial_r g_{\rho r} - \partial_\rho g_{rr}\right).\tag{4.40}$$

The metric $g^{r\rho}$ is non-null only for $\rho = r$, which leads to

$$\Gamma^r_{rr} = \frac{1}{2\tilde{g}}\partial_r\tilde{g}.\tag{4.41}$$

In the same way we find

$$\Gamma^t_{rt} = \frac{1}{2}g^{tt}\partial_r g_{tt} = \frac{1}{2}\frac{1}{\tilde{f}}\partial_r\tilde{f} \qquad \Gamma^t_{rr} = 0 \qquad \Gamma^x_{rx} = \Gamma^y_{ry} = \frac{1}{2}\frac{1}{\tilde{h}}\partial_r\tilde{h},\tag{4.42}$$

$$\Gamma^r_{xx} = \Gamma^r_{yy} = -\frac{1}{2\tilde{g}}\partial_r\tilde{h} \qquad \Gamma^t_{tr} = \frac{1}{2\tilde{f}}\partial_r\tilde{f},\tag{4.43}$$

and so on. From equation (3.7) we get the Ricci tensor. For instance

$$R_{rr} = R^r_{rrr} + R^t_{rtr} + R^x_{rxr} + R^y_{ryr},\qquad(4.44)$$

and from equation (3.9) we obtain $R^\rho_{\sigma\mu\nu}$. I will give two examples:

$$R^r_{rrr} = \partial_r\Gamma^r_{rr} - \partial_r\Gamma^r_{rr} + \Gamma^r_{r\mu}\Gamma^\mu_{rr} - \Gamma^r_{r\mu}\Gamma^\mu_{rr} = 0,\qquad(4.45)$$

$$R^t_{rtr} = -\partial_r\Gamma^t_{tr} + \Gamma^t_{tr}\Gamma^r_{rr}.\qquad(4.46)$$

We remark that in this text only repeated Greek symbols imply summation. Putting all the results together we obtain:

$$R_{rr} = -\frac{1}{2}\frac{\mathrm{d}}{\mathrm{d}r}\left(\frac{r^2}{f}\frac{\mathrm{d}}{\mathrm{d}r}\frac{f}{r^2}\right) + \frac{1}{4}\frac{fg}{r^4}\frac{\mathrm{d}}{\mathrm{d}r}\left(\frac{f}{r^2}\right)\frac{\mathrm{d}}{\mathrm{d}r}\left(\frac{g}{r^2}\right).\qquad(4.47)$$

Taking the above result into equation (3.14), $R_{rr} = -dg_{rr}$, we solve the differential equation. We follow the same procedure for the other components. The calculation is lengthy but straightforward. The final result is

$$\mathrm{d}s^2 = \frac{L^2}{r^2}\left(-f(r)\mathrm{d}t^2 + \frac{\mathrm{d}r^2}{f(r)} + \mathrm{d}x^i\mathrm{d}x^i\right),\qquad(4.48)$$

with

$$f(r) = 1 - \left(\frac{r}{r_+}\right)^d,\qquad(4.49)$$

where r_+ is an integration constant, which we will interpret below. Since $f \to 1$ as $r \to 0$, this space time is asymptotically AdS.

A remarkable feature of the Einstein equation is that for empty space it gives, besides the trivial Minkowski solution, a non-trivial solution which is interpreted as a black hole (see appendix A).

Before moving on, let us consider a metric written as $\mathrm{d}s^2 = -g_{tt}\mathrm{d}t^2 + g_{rr}\mathrm{d}r^2$. The light cones are given by the condition $\mathrm{d}s^2 = 0$, which leads to

$$\frac{\mathrm{d}t}{\mathrm{d}r} = \pm\sqrt{\frac{g_{rr}}{g_{tt}}}.\qquad(4.50)$$

If $\mathrm{d}t/\mathrm{d}r \to \pm\infty$ the light cones close up. Thus a light ray that approaches the singularity $r = r_+$ never seems to get there, at least in the above coordinate system. So there is a horizon at $r = r_+$. The surface at $r = r_+$ is infinitely red-shifted with respect to an asymptotic observer. This solution is generally known by the name 'black hole' (see appendix A), although in our case it would be better known as a 'black membrane'. Events at $r > r_+$ cannot influence the boundary near $r = 0$.

As is well known the partition function at finite temperature T is expressed as a Euclidean path integral over a periodic Euclidean time path. Thus finite temperature field theory is obtained by considering periodic imaginary time. We can use this approach to deduce that black holes radiate thermally at a given temperature T, called the Hawking temperature.

To proceed we perform an analytic continuation to Euclidean time $\tau = it$ and write the metric as

$$ds_E^2 = \frac{L^2}{r^2}\left(f(r)d\tau^2 + \frac{dr^2}{f(r)} + dx^i dx^i\right). \tag{4.51}$$

Near $r = r_+$ we can write

$$f(r) \approx f(r_+) + (r - r_+)f'(r_+) = (r - r_+)f'(r_+), \tag{4.52}$$

and the metric takes the following form in the vicinity of the horizon

$$ds_E^2 \approx \frac{f'(r_+)(r - r_+)}{r_+^2}d\tau^2 + \frac{dr^2}{r_+^2 f'(r_+)(r - r_+)} + \frac{dx^i dx^i}{r_+^2}. \tag{4.53}$$

This metric has a coordinate singularity at the horizon, which we can remove by a suitable choice of coordinates [11]. That is, at $r = r_+$ the Euclidean time direction shrinks to a point. This is similar to what happens at the origin of polar coordinates

$$ds_2 = d\rho^2 + \rho^2 d\varphi^2, \tag{4.54}$$

where φ is the polar angle. At $\rho = 0$ the metric is singular, but we know that the geometry is regular provided φ has period 2π. If $\varphi \in [0, 2\pi - \delta]$ with $\delta \neq 0$, $\rho = 0$ is a singular point, and the metric describes a cone. Changing coordinates so that the $\{r, t\}$ part of the Euclidean metric looks like the $\{\rho, \varphi\}$ metric near the horizon we find

$$d\rho^2 = \frac{dr^2}{r_+^2 f'(r_+)(r - r_+)}, \tag{4.55}$$

$$\rho = \frac{2}{r_+\sqrt{f'(r_+)}}. \tag{4.56}$$

Writing $\varphi = \beta\tau$ we find $\beta = f'(r_+)/2$ where $|f'(r_+)| = d/r_+$.

This metric describes a regular geometry if φ is periodic with period 2π:

$$\beta\tau \rightarrow \beta\tau + 2\pi \qquad \tau \rightarrow \tau + \frac{2\pi}{\beta} = \tau + \frac{4\pi r_+}{d}. \tag{4.57}$$

Identifying the Hawking temperature as the inverse of the periodicity we get

$$T = \frac{d}{4\pi r_+}. \tag{4.58}$$

With a change of variable $u = 1/r$, the black hole radius is $u = 1/r_+$. The temperature (4.58) increases as the black hole radius grows along the radial direction.

Thus a QFT in the presence of a black hole has a temperature T. Generally, the specific heat of a black hole is negative and the black hole is thermodynamically unstable, but in the case of an AdS space the black hole is stable and therefore represents an equilibrium situation. Notice that at zero temperature $r_+ = \infty$, and we recover the AdS metric. The area of the horizon (the set of points with $r = r_+$ and t fixed) is

$$A = \int_\Sigma \sqrt{g}\, \mathrm{d}^{d-1}x = \left(\frac{L}{r_+}\right)^{d-1} V, \qquad (4.59)$$

where V is the spatial volume of the boundary Σ_{d-1} (Σ_{d-1} is some manifold with dimension $d-1$; we can give it finite volume as an IR regulator). The entropy is given by

$$S = \frac{A}{4G_N} = \frac{L^{d-1}V}{4G_N r_+^{d-1}} = \frac{L^{d-1}V}{4G_N}\left(\frac{4\pi T}{d}\right)^{d-1}. \qquad (4.60)$$

The entropy density $s\,(= S/V)$ is

$$s = \frac{a}{4G_N}, \qquad (4.61)$$

where $a \equiv A/V$ is the area per unit volume (area density). Calculation of the entropy for a strongly interacting QFT is generally difficult. Using AdS/CFT correspondence the entropy can be found simply through calculating the area of the horizon. Dissipation due to the finite temperature is described as matter falling through the black hole horizon (figure 4.2).

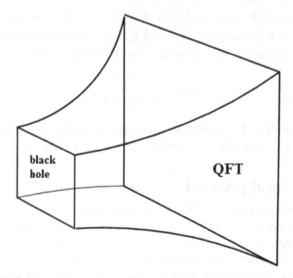

Figure 4.2. The QFT at finite temperature is in duality with classical gravity in a space time with a black hole. Waves propagating in the bulk become damped as they lose energy falling across the black hole. This damping is related to dissipation of the boundary theory [12].

The entropy is now calculated directly from the action. The on-shell gravity action for the black hole solution is given by

$$S = S_{\mathrm{E}} + S_{\mathrm{GH}} + S_{\mathrm{ct}}, \tag{4.62}$$

where

$$S_{\mathrm{E}} = -\frac{1}{16\pi G_{\mathrm{N}}} \int \mathrm{d}^{d+1}x \sqrt{g} \left[R + \frac{d(d+1)}{L^2} \right] \tag{4.63}$$

is the bulk gravity Euclidean action (see equation (3.15)), the second term is the Gibbons–Hawking boundary term (see equation (3.17)), and S_{ct} is an intrinsic boundary counter-term, needed to subtract divergences as $r \to 0$ and render the action finite, given by [8]:

$$S_{\mathrm{ct}} = \frac{1}{8\pi G_{\mathrm{N}}} \int_{\partial\Sigma} \mathrm{d}^d x \sqrt{\gamma} \frac{(d-1)}{L}. \tag{4.64}$$

The metric γ is defined by

$$\mathrm{d}s^2 \underset{r\to 0}{\approx} \frac{L^2}{r^2}\mathrm{d}r^2 + \gamma_{\mu\nu}\mathrm{d}x^\mu\mathrm{d}x^\nu. \tag{4.65}$$

Using $R = -(d+1)d/L^2$, we can write

$$S_{\mathrm{E}} = -\frac{1}{16\pi G_{\mathrm{N}}}\frac{d}{L^2}\int \mathrm{d}^{d+1}x \sqrt{g}. \tag{4.66}$$

We have shown in section 4.2 that $\sqrt{g} = (L/r)^{d+1}$. The integral in r is calculated from a cut off $r = \varepsilon$, up to $r = r_+$ and after that we integrate on a space with geometry $S^1 \times \Sigma_{d-1}$, where S^1 has the radius $1/T$, and Σ_{d-1} is a manifold with dimension $d-1$. The other terms are easily calculated at $r = \varepsilon$. The final result is

$$S = -\frac{L^{d-1}}{16\pi G_{\mathrm{N}}}\frac{V}{r_+^d T}, \tag{4.67}$$

where V is the volume of Σ_{d-1}. From equation (4.67) we obtain the free energy $F = TS$ and the entropy $S = -\partial F/\partial T$. We find the same result given by equation (4.60).

4.4 Finite chemical potential

In condensed matter we work with conserved charges: for instance, the number of electrons in a system does not change in time. The 'conserved charge' in the boundary has a dual in the bulk in the form of classical Maxwell electrodynamics. So the physics of a system characterized by a conservation law is described in the bulk by a classical Einstein equation (see equation (3.19)). In this section we will consider a black hole with charge. The charge of the black hole translates into the fact that the field theory is in a state with non-zero charge. A finite charge density is induced by holding the system at a non-zero chemical potential. The GKPW rule says that currents J_μ in the boundary are in duality with gauge fields in the bulk. From equation (3.15) and (3.19), we have the standard action for Einstein–Maxwell theory

$$S = \int d^4x \sqrt{g} \left[\frac{1}{2k^2} \left(R + \frac{6}{L^2} \right) - \frac{1}{4g_4^2} F_{\mu\nu} F^{\mu\nu} \right], \tag{4.68}$$

where g_4 is the Maxwell coupling, and $k^2 = 8\pi G_N$. In the full string theory context of gravity duality, there are other matter fields in addition to those in (4.68). If the Maxwell field is coupled to a scalar field, we will have to look for more general actions, as will be discussed in the conclusions. The Maxwell equation is given by equation (3.27). The Einstein equation of motion is now

$$R_{\mu\nu} - \frac{R}{2} g_{\mu\nu} - \frac{d(d-1)}{2L^2} g_{\mu\nu} = \frac{k^2}{2g_4^2} \left(2F_{\mu\eta} F^\eta_\nu - \frac{1}{2} g_{\mu\nu} F_{\alpha\beta} F^{\alpha\beta} \right). \tag{4.69}$$

Let us first consider the zero density state. For more detail see [9]. Examples of these systems are the 'zero chemical potential' superfluid Bose–Mott insulator transition and the 'interacting graphene' model. As pointed out by Zaanen et al [9], the first requirement is the presence of a continuous QPT. The quantum critical states realized at these transitions should however be strongly interacting in order to rely on universality. Only the systems with two space dimensions in condensed matter meet these requirements.

For zero density the Maxwell sector in equation (4.68) does not affect the AdS bulk and to calculate the optical conductivity we just need to calculate how infinitesimal electromagnetic perturbations propagate through the space time. The optical conductivity can be calculated using the following equation (see section 5.6)

$$\sigma(\omega) = -\frac{i}{\omega} G_{xx}^R(\omega, k = 0). \tag{4.70}$$

The matter in the boundary is neutral (does not have a charge) and the source and applied fields behave like chemical potential differences. To use the Maxwell equation, we need to choose a gauge. A convenient choice is $A_r = 0$ (since the components in the radial direction have no meaning in the boundary). The components A_x and A_y decouple, but A_x couples with A_t. However for $\vec{k} = 0$, A_x decouples from the other fluctuations, and only this component will be considered here. Writing $A_x = a_x(r) e^{-i\omega t}$ we have

$$F_{rx} = -F_{xr} = \frac{\partial a_x}{\partial r} e^{-i\omega t} \qquad F_{tx} = -F_{xt} = -i\omega a_x(r) e^{-i\omega t}. \tag{4.71}$$

Using now the Maxwell equation (3.27) with the metric (4.48) at $T = 0$, we obtain

$$\frac{\partial^2 a_x}{\partial r^2} + L^4 \omega^2 a = 0. \tag{4.72}$$

The solution of equation (4.72) is

$$a_x = c_0 e^{i(\omega L^2 r - \omega t)}, \tag{4.73}$$

where c_0 is a constant. For small r (near the boundary) we can write

$$A_x = A_x^{(0)} + A_x^{(1)} r + \cdots \tag{4.74}$$

where $A_x^{(0)} = c_0 e^{-i\omega t}$, and $A_x^{(1)} = i c_0 \omega L^2 e^{-i\omega t}$.

As was mentioned in section 4.1 the GKPW rule says that $A_x^{(0)}$ is the source in the boundary and $A_x^{(1)}$ is the corresponding response. The conductivity is given by equation (4.70). We find [9]:

$$\sigma(\omega) = -\frac{i A_x^{(1)}}{L^2 g_4^2 \omega A_x^{(0)}} = \frac{1}{g_4^2}. \tag{4.75}$$

To calculate the optical conductivity at finite temperature we can use the formalism described in chapter 5. The calculations are presented in [9]. Surprisingly, it was found that the result at zero temperature was unchanged at finite temperature.

Now we turn to finite densities (figure 4.3). We will consider a Maxwell field where we have only the scalar potential $A_t(r)$. Taking the metric (4.38) in equation (4.68) we find [2]

$$f(r) = 1 - \left(1 + \frac{r_+^2 \mu^2}{\gamma^2}\right)\left(\frac{r}{r_+}\right)^d + \frac{r_+^2 \mu^2}{\gamma^2}\left(\frac{r}{r_+}\right)^{2(d-1)}, \tag{4.76}$$

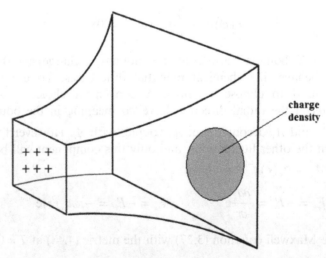

Figure 4.3. A black hole with charge. The charge density is sourced by the flux emanating from the black hole horizon (adapted from reference [1]).

where

$$\gamma^2 = \frac{(d-1)g_4^2 L^2}{(d-2)k^2},$$ (4.77)

and

$$A_t = \mu\left[1 - \left(\frac{r}{r_+}\right)^{d-2}\right].$$ (4.78)

Here we have imposed the condition that A_t vanishes on the horizon, in order for $A_\nu A^\nu$ to remain finite there. For $r \ll r_+$ we have $f(r) \approx 1$, and the metric becomes that of pure AdS. This means that the physics at energy scales much larger than the chemical potential μ is simply that of the vacuum. For $d = 3$, and measuring the distance from the horizon, we take $u = 1/r$, then from equation (4.78) the electric field is $\sim 1/u^2$. The surface area is $\sim u^2$, thus the total electric flux in the bulk is conserved between the horizon and the boundary [12]. The horizon is determined by the zero of the function $f(r)$, $f(r) = 0$. This function (unlike the case in section 4.3) allows for two solutions representing an inner and an outer horizon, but the inner horizon does not play any role. The above solution describes an electrically charged black hole, known as the Reissner–Nordström AdS black hole, and this is the dual of a d-dimensional strongly coupled field theory at finite temperature T and finite charge density. The asymptotic value of the bulk gauge field $A_t(0) = \mu$ is interpreted in the duality theory as the chemical potential for electric charge density.

The chemical potential μ is the energy needed to add one particle to a thermally and mechanically isolated system. For a particle with charge e in an electric potential A_t we have $\mu = eA_t$. We remark that as A_t depends only on r, the electric field is in the r direction in the bulk and the rotational symmetry is preserved.

The function $f(r)$ can also be written as

$$f(r) = 1 + Q^2 r^{2(d-1)} - Mr^d,$$ (4.79)

where

$$Q^2 = \frac{\mu^2}{\gamma^2 r_+^{2d-4}} \qquad M = \frac{1}{r_+^d} + Q^2 r_+^{d-2},$$ (4.80)

M is the mass and Q the charge of the black hole. The charge Q of the black hole is proportional to μ. The temperature can be calculated using the analytic continuation to a Euclidean metric, and we obtain

$$T = \frac{1}{4\pi r_+}\left[d - \frac{(d-2)r_+^2 \mu^2}{\gamma^2}\right],$$ (4.81)

or

$$T = \frac{1}{4\pi r_+}\left(3 - Q^2 r_+^4\right). \tag{4.82}$$

The parameter Q is bounded between $0 \leqslant Q \leqslant 3/r_+^4$, interpolating between AdS–Schwarzschild and the extremal AdS black hole. Due to the scale invariance of the system, the temperature by itself does not have any physical meaning but dimensionless combinations such as T/μ are meaningful.

If $T = 0$, equation (4.81) gives

$$r_+ = \left(\frac{d}{d-2}\right)^{\frac{1}{2(d-1)}} Q^{\frac{1}{1-d}}. \tag{4.83}$$

The black hole still exists with a finite horizon radius. This is called an 'extremal black hole', characterized by a mass that is entirely due to the electromagnetic charge. When the mass is entirely due to the electrical charge it is impossible to reduce it further by Hawking radiation. The entropy of this object is finite, since the horizon area is finite. Finite entropy at zero temperature implies a degenerate ground state. In condensed matter, frustrated systems have finite entropy at zero temperature. However the Reissner–Nordström state is different from the ground state of these frustrated systems. It is believed that when back reaction to the black hole metric is included, the black hole horizon is replaced by one with no zero temperature entropy. One example will be considered in section 7.1.

In the following we will consider the extremal black hole near the horizon. In this region it is more convenient to use $r = r_+ + \xi$. The metric (4.48) then becomes

$$ds^2 = L^2 \left(dx^2 + dy^2 - a\xi^2 dt^2 + \frac{d\xi^2}{a\xi^2}\right) \tag{4.84}$$

where $a = 6/r_+^2$. Making a change of variable $\eta = 1/\xi$, we obtain

$$ds^2 = L^2 \left[dx^2 + dy^2 + \frac{1}{\eta^2}\left(-\frac{6}{r_+^2}dt^2 + \frac{r_+^2 d\eta^2}{6}\right)\right]. \tag{4.85}$$

The space factorizes into a two-dimensional anti-de Sitter space (AdS$_2$) in r and t, and a flat R_2 in the spatial coordinates of the boundary. The effective AdS geometry is realized in the time-radial direction plane. The scaling isometry of the metric is now

$$t \to \lambda t \qquad \eta \to \lambda \eta \qquad x \to x. \tag{4.86}$$

As pointed out by Faulkner et al [13], little is known about AdS$_2$/CFT$_1$ duality. It is not clear whether the dual theory is conformal quantum mechanics or a chiral sector of a $(1+1)$-dimensional CFT.

In the spatial direction there is no scale invariance; one is dealing with a purely temporal quantum criticality. This change of space near the horizon implies critical scaling in frequency alone, and so marginal Fermi liquid behavior of the boundary propagators, with self-energy given by $\Sigma(\vec{\mathbf{k}}, \omega) \propto \omega \ln \omega$. Such emergent scale invariance is ubiquitous in the strange metals [14] (as we will see in section 7.2), and this phenomenon is unknown in bosonic field theory. The AdS_2 metric may be thought of as encoding local quantum criticality by imposing scaling only in the temporal dimension.

A background magnetic field $B = F_{(0)xy}$ preserves rotational symmetry only in a $(2 + 1)$-dimension field theory. Adding a magnetic field in the case $d = 3$ is straightforward. One finds a dyonic black hole, i.e. a black hole with both electric and magnetic charge [8].

The metric is again given by equation (4.48), but now

$$f(r) = 1 - \left[1 + \frac{\left(r_+^2\mu^2 + r_+^4 B^2\right)}{\gamma_2}\right]\left(\frac{r}{r_+}\right)^3 + \frac{\left(r_+^2\mu^2 + r_+^4 B^2\right)}{\gamma^2}\left(\frac{r}{r_+}\right)^4, \qquad (4.87)$$

and

$$A_t = \mu\left(1 - \frac{r}{r_+}\right) \qquad A_y = Bx. \qquad (4.88)$$

The temperature will be given by

$$T = \frac{1}{4\pi r_+}\left(3 - \frac{r_+^2\mu^2}{\gamma^2} - \frac{r_+^4 B^2}{\gamma^2}\right). \qquad (4.89)$$

To avoid the non-zero ground state entropy problem, other metrics beyond Reissner–Nordström geometry have been studied. I will not go into detail here but refer the reader to references [1, 12]. One of these metrics has a different form from those studied here. It no longer proposes a horizon and has different scalings in both space and time. This metric has been called the 'Lifshitz' metric. However, a physical interpretation of the results obtained using this metric remains unclear.

References

[1] Hartnoll S A Horizons, holography and condensed matter arXiv:1106.4324 [hep-th]
[2] McGreevy J Holographic duality with a view toward many-body physics arXiv:0909.05182 [hep-th]
[3] Altland A and Simons B 2007 *Condensed Matter Field Theory* (Cambridge: Cambridge University Press)
[4] Maldacena J M 1998 The large N limit of superconformal field theories and supergravity *Adv. Theor. Math. Phys.* **2** 231
Klebanov I R and Maldacena J M 2009 Solving quantum field theories via curved space-times *Physics Today* January 28
Johnson C V and Steiberg P 2010 What black holes teach about strongly coupled particles *Physics Today* May 29

Nastase H Introduction to AdS-CFT arXiv:0712.06889 [hep-th]

Aharony O, Gubser S S, Maldacena J M, Ooguri H and Oz Y 2000 Large N field theories, string theory and gravity *Phys. Rep.* **323** 183

[5] Gubser S S, Klebanov I R and Polyakov A M 1998 Gauge theory correlators from non-critical string theory *Phys. Lett.* B **428** 105

[6] Witten E 1998 Anti-de Sitter space and holography *Adv. Theor. Mat. Phys.* **2** 253

[7] Hartnoll S A Quantum critical dynamics from black holes arXiv:0909.3553 [cond-mat.str-el]

[8] Hartnoll S A 2009 Lectures on holographic methods for condensed matter physics *Class. Quantum Grav.* **26** 224002

[9] Zaanen J, Sun Y W, Liu Y and Schalm K The AdS/CFT manual for plumbers and electricians www.lorentz.leidenuniv.nl/~kschalm/papers/adscmtreview.pdf

[10] Sachdev S 1999 *Quantum Phase Transitions* (Cambridge: Cambridge University Press)

[11] Petersen J L 1999 Introduction to the Maldacena conjecture on AdS/CFT *Int. J. Mod. Phys.* A **14** 3597

[12] Sachdev S 2012 What can gauge-gravity duality teach us about condensed matter physics *Annu. Rev. Cond. Matt. Phys.* **3** 9

[13] Faulkner T, Iqbal N, Liu H, McGreevy J and Vegh D From black holes to strange metals arXiv:1003.1728 [hep-th]

[14] Green A G An introduction to gauge gravity duality and its application in condensed matter arXiv:1304.5908 [cond-mat]

Chapter 5

Dynamics

5.1 Linear response theory

In this section I will present a brief review of the linear response theory, which provides a general framework for studying the dynamical properties of a condensed matter system close to thermal equilibrium. Most of the experiments in physics are conveniently discussed in terms of correlation functions. In statistical systems which are near thermal equilibrium they give all the desirable information about the intrinsic statistical fluctuations. They afford the most convenient vehicles in terms of which one bridges the gap between a microscopic reversible description and the irreversible macroscopic behavior of many body systems. In an experiment in condensed matter, the physicist usually uses as a probe an external force which disturbs the system slightly from equilibrium, and then measures the linear response to this force. In fact what is measured is the dynamical behavior of spontaneous fluctuations about the equilibrium state. These fluctuations can be rigorously described in terms of time dependent correlation functions [1].

To start we consider a system at equilibrium described by a Hamiltonian H_0, subjected to an external time dependent force $F(t)$, leading to a perturbation term

$$H'(t) = -AF(t), \qquad (5.1)$$

where A is a dynamical variable coupled to F. For instance, F could represent a time dependent magnetic field coupled to the magnetization $A = M$. We will consider weak perturbation and ask for the response of the system in a linear approximation. Sometimes the operator associated with the response is designated by O. Here, for simplicity, I will consider only operators that do not depend on position. The external force will produce a change in the expectation value of another dynamical variable B (in some cases $B = A$). The expectation value of B is given by

$$\langle B(t) \rangle = tr \rho(t) B, \qquad (5.2)$$

doi:10.1088/978-1-627-05309-9ch5   5-1

where ρ is the time dependent density matrix $\rho = e^{-\beta H}/Z$, with $Z = tre^{-\beta H}$ and $tr\rho(t) = 1$. We write $\rho(t)$ as

$$\rho(t) = \rho_0 + \Delta\rho, \qquad (5.3)$$

with $\rho(t = -\infty) = \rho_0$, where $\rho_0 = e^{-\beta H_0}/Z$.

The equation of motion for ρ is given by

$$i\hbar\frac{\partial\rho(t)}{\partial t} = [H, \rho(t)] = [H_0, \rho(t)] + [H', \rho(t)]. \qquad (5.4)$$

Neglecting the term $\Delta\rho H'$ we have

$$i\Delta\dot{\rho}(t) \approx [H_0, \Delta\rho(t)] + \left[H'(t), \rho_0\right], \qquad (5.5)$$

which can be written in the form

$$\Delta\rho(t) = -i\int_{-\infty}^{t} e^{-iH_0(t-t')}\left[H'(t'), \rho_0\right]e^{iH_0(t-t')}dt'. \qquad (5.6)$$

The induced change in the variable B from its equilibrium value is

$$\Delta B(t) = tr(\rho B) = tr\left[\left(\rho_0 + \Delta\rho\right)B\right] = \langle B\rangle_0 + tr(\Delta\rho B). \qquad (5.7)$$

Setting $\langle B\rangle_0 = 0$, we get

$$\Delta B(t) = itr\int_{-\infty}^{t} e^{-iH_0(t-t')}\left[A, \rho_0\right]e^{iH_0(t-t')}BF(t')dt'. \qquad (5.8)$$

Using $tr(AB) = tr(BA)$ we obtain

$$\Delta B(t) = itr\int_{-\infty}^{t} dt'\left[A, \rho_0\right]B(t - t')F(t'), \qquad (5.9)$$

where $B(t) = e^{-H_0 t}Be^{-iH_0 t}$. We can write equation (5.9) as

$$\Delta B(t) = i\int_{-\infty}^{t} \langle[B(t - t'), A]\rangle F(t')dt'. \qquad (5.10)$$

Defining the linear response function as

$$K(t) = i\langle[B(t), A]\rangle, \qquad (5.11)$$

which describes how the system responds to the application of F, we have

$$\Delta B(t) = \int_{-\infty}^{t} dt'K(t - t')F(t'). \qquad (5.12)$$

If $F(t)$ is of the form $F(t) = e^{-i\omega t}$ we define a dynamical susceptibility $\chi(\omega)$ by

$$\Delta B(t) = \chi(\omega)F(t), \qquad (5.13)$$

then

$$\Delta B(t) = \int_{-\infty}^{t} dt' K(t - t') e^{-i\omega t'} F. \tag{5.14}$$

Making a change of variable $t - t' = \tau$ and then $t = \tau$ we get for $\chi(\omega)$

$$\chi(\omega) = \int_{0}^{\infty} e^{i\omega t} K(t) dt. \tag{5.15}$$

To make sure that the integrals converge we should have written $F(t) = e^{-i\omega t + \epsilon t}$ and then taken the limit $\epsilon \to 0$. This procedure will be always implicit. The susceptibility of the order parameter is an important observable to characterize the dynamical nature of a QPT. At the onset of instability (i.e. a critical point) the static susceptibility diverges, reflecting the tendency of the system to develop an expectation value of the operator O (associated with the response) even in the absence of an external source. The Laplace transform of $K(t)$ is given by

$$\tilde{\chi}(z) = \int_{0}^{\infty} dt e^{izt} K(t) \qquad \text{for } \Im z > 0, \tag{5.16}$$

where $\Im$ means the imaginary part. The susceptibility $\chi(\omega)$ is the limit of $\tilde{\chi}(z)$ as we approach the real frequency axis from above

$$\chi(\omega) = \lim_{\epsilon \to 0} \tilde{\chi}(\omega + i\epsilon) = \chi'(\omega) + i\chi''(\omega). \tag{5.17}$$

If the system is translationally invariant in time we can define a Fourier transform by

$$K(t) = \frac{1}{2\pi} \int_{-\infty}^{\infty} e^{-i\omega t} K(\omega) d\omega, \tag{5.18}$$

then

$$\tilde{\chi}(z) = \frac{1}{2\pi} \int_{0}^{\infty} dt \int_{-\infty}^{\infty} e^{-i(\omega - z)t} K(\omega) d\omega. \tag{5.19}$$

Integrating in t we find

$$\tilde{\chi}(z) = \frac{1}{2\pi i} \int_{-\infty}^{\infty} \frac{K(\omega) d\omega}{\omega - z}. \tag{5.20}$$

Using the identity

$$\frac{1}{x \pm i0} = P\frac{1}{x} \mp i\pi\delta(x), \tag{5.21}$$

where P denotes the principal value, we obtain

$$\tilde{\chi}(\omega + i\epsilon) = P\frac{1}{2\pi} \int_{-\infty}^{\infty} \frac{K(\omega') d\omega'}{\omega' - \omega} + K(\omega). \tag{5.22}$$

Thus

$$\chi''(t) = \frac{1}{2} \langle [B(t), A] \rangle, \tag{5.23}$$

and

$$\tilde{\chi}(z) = \frac{1}{\pi} \int_{-\infty}^{\infty} \frac{\chi''(\omega) d\omega}{\omega - z}. \tag{5.24}$$

The reader should be aware that the words perturbation theory are used in two different contexts. We have perturbation theory, mainly in terms of Feynman diagrams, to treat a many body interacting system such as the electrons in a metal. For a strongly correlated system it may not work. On the other hand, we can use the perturbation technique in linear response theory (as was done above). Here the perturbation is caused by an external agent (as an experimental probe) and can always be made small.

5.2 Relation between susceptibility and Green's function

The retarded Green's function for two observables A and B is given in terms of the expectation value for their commutators as follows:

$$G_{BA}^R(t - t') = \langle\langle B(t - t'), A \rangle\rangle = -i\theta(t - t')\langle [B(t - t'), A] \rangle, \tag{5.25}$$

where $\theta(t)$ is the Heaviside step function; non-zero and equal to one for $t > 0$. For fermion operators we use an anticommutator in (5.25). We see that the retarded Green's function is causal. That is, the expectation value at time t depends only on the source at times $t' < t$. The retarded Green's function carries all the real time information about the response of the system, it can be obtained directly from a path integral formulation, and is very convenient to calculate in perturbation theory; but on the other hand it can be related, by suitable analytic continuation, to the quantities arising from experimental observables. As will be discussed in section 7.2, G^R for a fermionic operator can be used to probe the existence of a Fermi surface and the nature of the low energy excitations around this surface. Equation (5.10) can be written as

$$\Delta B(t) = itr \int_{-\infty}^{t} \rho_0 [B(t - t')] F(t') dt'$$
$$= itr \int_{-\infty}^{\infty} \theta(t - t') [B(t - t'), A] F(t') dt', \tag{5.26}$$

or

$$\Delta B(t) = -\int_{-\infty}^{\infty} G_{BA}^R(t - t') F(t') dt'. \tag{5.27}$$

If $F(t) = F e^{-i\omega t}$ we get

$$\chi(\omega) = -\int_{-\infty}^{\infty} G_{BA}^R(t - t') e^{i\omega(t - t')} dt', \tag{5.28}$$

which leads to the relation between susceptibility and Green's function

$$\chi(\omega) = -2\pi G_{BA}^R(\omega). \tag{5.29}$$

The inverse Fourier transform of the Green's function

$$G_{AB}^R(t) = \int_{-\infty}^{\infty} \frac{d\omega}{2\pi} e^{-i\omega t} G_{AB}^R(\omega), \tag{5.30}$$

can be evaluated for $t < 0$ by closing the ω contour in the upper half-plane. Causality implies that we must obtain a null result for $G^R(t)$ at $t < 0$. Therefore $G_{AB}^R(\omega)$ is analytic in ω, for $\Im \omega > 0$. This analyticity property leads to the well-known Kramers–Kronig relation between the real and imaginary parts of G^R:

$$\Re G^R(\omega) = P \int_{-\infty}^{\infty} \frac{d\omega'}{\pi} \frac{\Im G^R(\omega)}{\omega' - \omega}, \tag{5.31}$$

$$\Im G^R(\omega) = -P \int_{-\infty}^{\infty} \frac{d\omega'}{\pi} \frac{\Re G^R(\omega)}{\omega' - \omega}. \tag{5.32}$$

5.3 The relaxation function

We have from (5.11) and (5.15)

$$\chi_{BA}(\omega) = i \int_0^{\infty} \langle [B(t), A] \rangle e^{i\omega t} dt, \tag{5.33}$$

which can be written as

$$\chi_{BA}(\omega) = i tr \int_0^{\infty} [A, \rho_0] B(t) e^{i\omega t} dt. \tag{5.34}$$

For any operator A, the following identity may be verified by direct differentiation with respect to an auxiliary variable λ:

$$\frac{d}{d\lambda} \{ e^{\lambda H_0} [A, e^{-\lambda H_0}] \} = H_0 e^{\lambda H_0} A e^{-\lambda H_0} - e^{\lambda H_0} A e^{-\lambda H_0} H_0 = [H_0, A(-i\lambda)]$$

$$= -i \frac{d}{dt} A(-i\lambda), \tag{5.35}$$

where $A(-i\lambda) = e^{\lambda H_0} A e^{-\lambda H_0}$. Integrating with respect to λ up to the value β we obtain

$$[A, \rho_0] = -i\rho_0 \int_0^{\beta} \frac{d}{dt} A(-i\lambda) d\lambda. \tag{5.36}$$

From (5.34) and (5.36) we have

$$\chi_{BA}(\omega) = \int_0^{\infty} e^{i\omega t} \varphi_{BA}(t) dt, \tag{5.37}$$

where

$$\varphi_{BA}(t) = tr \int_0^\beta \rho_0 \frac{dA(-i\lambda)}{dt} B(t) d\lambda = -tr \int_0^\beta \rho_0 A(-i\lambda) \frac{dB(t)}{dt} d\lambda. \qquad (5.38)$$

The dynamical susceptibility can then be written as

$$\chi_{BA}(\omega) = -tr\rho_0 \int_0^\beta A(-i\lambda) \left(\int_0^\infty dt e^{i\omega t} \frac{dB}{dt} dt \right) d\lambda. \qquad (5.39)$$

Integrating by parts we get

$$\chi_{BA}(\omega) = tr\rho_0 \int_0^\beta A(-i\lambda) \left[B(0) - i\omega \int_0^\infty e^{-\omega t} B(t) dt \right] d\lambda. \qquad (5.40)$$

Let us define the relaxation function $R(t)$ by

$$R(t) = (A, B(t)) = \int_0^\beta \langle A(-i\lambda) B(t) \rangle d\lambda. \qquad (5.41)$$

then we can write

$$\chi_{AB}(\omega) = \chi_{AB}(0) \left[1 - i\omega \int_0^\infty \frac{R(t)}{\chi(0)} dt \right], \qquad (5.42)$$

where $\chi_{AB}(0) = (A, B)$.

The relaxation function is a very important theoretical quantity, since it can be calculated using the memory function formalism described in reference [1].

5.4 The fluctuation dissipation theorem

Let us write the correlation function as

$$S_{AB}(t) \equiv \langle A(t) B(0) \rangle - \langle A(t) \rangle \langle B(0) \rangle, \qquad (5.43)$$

and the imaginary part of the susceptibility

$$\chi''_{AB}(t) = \frac{1}{2} \langle [A(t), B] \rangle. \qquad (5.44)$$

We have subtracted the equilibrium averages in (5.43) so that $S_{AB}(t) \to 0$ as $t \to \infty$, and thus its Fourier transform $S_{AB}(\omega)$ is well defined. Of course $\langle A(t) \rangle$ is independent of time. Once more the attention of the reader is drawn to the different use of the word correlation function (see section 4.2). Here I am using the terminology used in condensed matter and statistical physics.

Because the operator $e^{-\beta H}$ causes a time translation by the imaginary time $\tau = i\beta$, we have

$$tre^{-\beta H} A(t) B(0) = trA(t + i\beta) e^{-\beta H} B(0) = tre^{-\beta H} B(0) A(t + i\beta), \qquad (5.45)$$

where we have used $trAB = trBA$. Time translation invariance implies $\langle A(t)B(0)\rangle = \langle A(0)B(-t)\rangle$, which leads to

$$S_{BA}(-t) = \langle B(-t)A(0)\rangle = \langle B(0)A(t)\rangle = tre^{-\beta H}A(0)B(-t + i\beta)$$
$$= \langle A(t - i\beta)B(0)\rangle, \tag{5.46}$$

or

$$S_{BA}(-t) = S_{AB}(t - i\beta) = e^{-i\beta\partial_t}S_{AB}(t). \tag{5.47}$$

From this equation we obtain

$$S_{BA}(-\omega) = e^{-i\beta\omega}S_{AB}(\omega), \tag{5.48}$$

which implies

$$\langle A(t)B(t')\rangle = \int_{-\infty}^{\infty} S_{AB}(\omega)e^{\beta\omega}e^{-i\omega(t-t')}d\omega. \tag{5.49}$$

From (5.44) we have

$$2\chi''_{AB}(t) = S_{AB}(t) - S_{BA}(-t) = [1 - e^{-i\beta\partial_t}]S_{AB}(t). \tag{5.50}$$

Fourier transforming this equation (using $\partial_t \rightarrow \omega$) we find

$$\chi''_{AB}(\omega) = \frac{1}{2}(1 - e^{-\beta\omega})S_{AB}(\omega). \tag{5.51}$$

This result is known as the *Fluctuation Dissipation Theorem*. It relates, for a system in thermal equilibrium, two physically distinct quantities: the spontaneous fluctuations on one side, described by $S_{AB}(t)$, which arise, even in the absence of external forces from the thermal motion of particles of the system, to the dissipative behavior on the other side, described by χ''.

From (5.41) this is written as

$$(A, B(t)) = tr\rho_0 \int_0^\beta e^{\lambda H_0}Ae^{-\lambda H_0}e^{iH_0t}Be^{-iH_0t}d\lambda, \tag{5.52}$$

and we see that

$$(A, B(-t)) = (A, B(t))^*, \tag{5.53}$$

where the star means complex conjugation. Using this relation we arrive at

$$R_{AB}(\omega) = \int_{-\infty}^{\infty} e^{-i\omega t}(A, B(t))dt = \int_{-\infty}^{0} e^{-i\omega t}(A, B(t))dt + \int_{0}^{\infty} e^{-i\omega t}(A, B(t))dt$$
$$= \int_{0}^{\infty} e^{i\omega t}(A, B(-t))dt + \int_{0}^{\infty} e^{-i\omega t}(A, B(t))dt$$
$$= \int_{0}^{\infty} e^{i\omega t}(A, B(t))^*dt + \int_{0}^{\infty} e^{-i\omega t}(A, B(t))dt$$
$$= 2\,\Re \int_{0}^{\infty} e^{-i\omega t}(A, B(t))dt. \tag{5.54}$$

Using (5.42) we obtain

$$\chi''_{AB}(\omega) = \frac{\omega}{2} R_{AB}(\omega). \tag{5.55}$$

Therefore, the fluctuation dissipation theorem can also be written as

$$R_{AB}(\omega) = \frac{(1 - e^{-\beta\omega})}{\omega} S_{AB}(\omega). \tag{5.56}$$

5.5 Properties of the Green's function

If $|n\rangle$ is the energy eigenstates of the system with eigenvalues E_n, the density operator can be written as

$$\rho = Z^{-1} \sum_n e^{-\beta E_n} |n\rangle\langle n|. \tag{5.57}$$

We assume the states $|n\rangle$ form a complete orthonormal system

$$\sum_n |n\rangle\langle n| = 1 \qquad \langle n||m\rangle = \delta_{nm}. \tag{5.58}$$

In this representation the correlation function takes the form

$$\begin{aligned}
S_{AB}(t) &= tr\left(Z^{-1} \sum_j e^{-\beta E_j} |j\rangle\langle j| A(t) B \right) \\
&= Z^{-1} \sum_{n,m,j} \langle n| e^{-\beta E_j} |j\rangle\langle j| e^{iHt} A e^{-iHt} |m\rangle\langle m| B |n\rangle \\
&= Z^{-1} \sum_{m,n} e^{-\beta E_n} A_{nm} B_{mn} e^{-iE_{mn}t},
\end{aligned} \tag{5.59}$$

where $E_{mn} = E_m - E_n$. The Fourier transform is given by

$$S_{AB}(\omega) = Z^{-1} \sum_{m,n} e^{-\beta E_n} A_{nm} B_{mn} \delta(\omega - E_{mn}), \tag{5.60}$$

where we have used

$$\frac{1}{2\pi} \int_{-\infty}^{\infty} e^{ixt} dx = \delta(t). \tag{5.61}$$

The argument of the delta function contains possible excitation energies for the system. From the retarded Green's function written as

$$G_{AB}^R(E) = \frac{1}{2\pi i} \int_{-\infty}^{\infty} dt e^{iE(t-t')} \theta(t - t') \langle A(t) B(t') - B(t') A(t) \rangle \tag{5.62}$$

and using (5.49) we can write

$$G^R_{AB}(E) = \frac{1}{2\pi i} \int_{-\infty}^{\infty} d\omega S_{AB}(\omega)(e^{\beta\omega} - 1) \int_{-\infty}^{\infty} dt e^{i(E-\omega)t} \theta(t). \qquad (5.63)$$

Using the following representation of the $\theta(t)$ function

$$\theta(t) = \frac{i}{2\pi} \int_{-\infty}^{\infty} dx \frac{e^{-ixt}}{x + i\varepsilon} \qquad \text{with } \varepsilon \to +0, \qquad (5.64)$$

we can write the last integral in (5.63) as

$$I = \int_{-\infty}^{\infty} dt e^{i(E-\omega)t} \theta(t) = \frac{i}{2\pi} \int_{-\infty}^{\infty} \int_{-\infty}^{\infty} \frac{e^{i(E-\omega-x)t}}{x + i\varepsilon} dx dt, \qquad (5.65)$$

where we have taken the variable x as complex. For $t > 0$, we close the contour (in the x-plane) in the lower half-plane. Since the integrand has a pole at $x = -i\varepsilon$, we get the residue

$$2\pi i \left(\frac{i}{2\pi} \right) e^{i(E-\omega+i\varepsilon)t}. \qquad (5.66)$$

If $t < 0$, we close the contour in the upper half-plane where the integrand has no singularities and the integral vanishes. Thus

$$I = -\int_{0}^{\infty} dt e^{i(E-\omega)t - \varepsilon t} = \frac{i}{E - \omega + i\varepsilon}. \qquad (5.67)$$

As a result, we get the spectral representation of the retarded Green's function

$$G^R_{AB}(E) = \frac{1}{2\pi} \int_{-\infty}^{\infty} (e^{\beta\omega} - 1) S_{AB}(\omega) \frac{d\omega}{E - \omega + i\varepsilon}. \qquad (5.68)$$

We can define an advanced Green's function by changing $\theta(t - t')$ to $-\theta(t' - t)$, which leads to

$$G^A_{AB}(E) = \frac{1}{2\pi} \int_{-\infty}^{\infty} (e^{\beta\omega} - 1) S_{AB}(\omega) \frac{d\omega}{E - \omega - i\varepsilon}. \qquad (5.69)$$

Considering E as a complex variable, we can combine (5.68) and (5.69) to write

$$G_{AB}(E) = \frac{1}{2\pi} \int_{-\infty}^{\infty} (e^{\beta\omega} - 1) S_{AB}(\omega) \frac{d\omega}{E - \omega} = \begin{cases} G^R_{AB}(E) & \text{for } \Im E > 0 \\ G^A_{AB}(E) & \text{for } \Im E < 0 \end{cases} \qquad (5.70)$$

From the Green's function we can obtain the spectral function as follows:

$$G_{AB}(\omega + i\varepsilon) - G_{AB}(\omega - i\varepsilon)$$

$$= \frac{1}{2\pi} \int_{-\infty}^{\infty} (e^{\beta\omega'} - 1) S_{AB}(\omega') \left[\frac{1}{\omega - \omega' + i\varepsilon} - \frac{1}{\omega - \omega' - i\varepsilon} \right] d\omega'$$

$$= -i \int_{-\infty}^{\infty} (e^{\beta\omega'} - 1) S_{AB}(\omega') \delta(\omega - \omega') d\omega'. \qquad (5.71)$$

which leads to

$$S_{AB}(\omega) = \frac{i}{e^{\beta\omega} - 1}\left[G_{AB}(\omega + i\varepsilon) - G_{AB}(\omega - i\varepsilon)\right]. \qquad (5.72)$$

The functions $G_{AB}^{R}(E)$ and $G_{AB}^{A}(E)$ are analytic in the upper and lower half-planes, respectively. G^{R} can be continued analytically into the upper and G^{A} into the lower complex planes, respectively. They can therefore be considered as two branches of a single analytic function $G_{AB}(E)$, having singularities along the real axis (poles or lines). The singularities of the Green's function give the energies of elementary excitation for the system. We can see this as follows. Suppose that a Green's function has poles only at points ω_n of the real axis. From equation (5.71) we see that the spectral function will have delta function type singularities at the points ω_n, and the correlation function (5.43) will oscillate at those frequencies. At zero temperatures, equation (5.60) implies that the quantities ω_n are exact eigenvalues of the Hamiltonian describing the system. For temperatures above zero, these quantities are functions of temperature and the chemical potential, and they cannot be given a purely mechanical meaning. Nevertheless, even in this case we can interpret them as undamped oscillations of the system.

More complicated singularities may appear on the real axis, but in this case we can find a sheet of the analytic function $G(E)$ upon which the singularities shift to the complex plane in the form of poles. The imaginary parts of the poles represent the damping of the corresponding oscillations. If the damping is weak, we can introduce the concept of quasi-particle states.

The strongly coupled theories which are of interest in the present text generally do not have quasi-particle excitations. Therefore, the presence of sharp features in the response function is not expected along the real frequency axis. The problem then is how to characterize the spectral functions in such theories. As pointed out by Hartnoll [2], in all the known retarded Green's functions obtained via the duality principle, the only non-analyticities found at finite temperature are poles at specific complex frequencies. These poles are called *quasi-normal frequencies* of the Green's function, and are denoted by ω^*. From equation (4.36) we see that the poles occur when $A(\omega^*, k) = 0$.

5.6 Transport properties

In this section I will present a short discussion about transport. Following [3, 4] spin transport in insulating magnets is analyzed, where spin currents can be generated by a magnetic field gradient. The same formalism can be used to study the transport of electric charges [5]. The final result is the same, but the calculations are simple for spin transport.

Let us consider a variable S with a current density operator $\vec{j}$ defined via the continuity equation

$$\frac{\partial S}{\partial t} + \vec{\nabla} \cdot \vec{j} = 0. \qquad (5.73)$$

Fourier transforming this equation we obtain

$$\frac{\partial S(\vec{q}, t)}{\partial t} + i\vec{q} \cdot \vec{j}(\vec{q}, t) = 0. \tag{5.74}$$

The linear current response to an external magnetic field gradient h along the z direction is given by

$$\langle j(\vec{q}, \omega) \rangle = \chi_{jS}(\vec{q}, \omega) h(\vec{q}, \omega). \tag{5.75}$$

For simplicity I will consider a current only in the x direction. The dynamic susceptibility is

$$\chi_{jS}(\vec{q}, \omega) = i \int_0^\infty dt e^{i\omega t} \langle [j(\vec{q}, t), S(-\vec{q}, 0)] \rangle. \tag{5.76}$$

Using equation (5.75) and performing a partial integration we obtain

$$\chi_{jS}(\vec{q}, \omega) = \frac{i}{\omega} \left\{ -\langle [j(\vec{q}), S(-\vec{q})] \rangle \right.$$
$$\left. + \int_0^\infty dt e^{i\omega t} \langle [j(\vec{q}, t), iq_x j(-\vec{q}, 0)] \rangle \right\}. \tag{5.77}$$

We can write equation (5.75) as

$$\langle j(\vec{q}, \omega) \rangle \frac{1}{i\omega} (\langle K \rangle + \Lambda(\vec{q}, \omega)) iq_x h(\vec{q}, \omega), \tag{5.78}$$

where

$$\langle K \rangle = \langle [j_x(\vec{q}), S(-\vec{q})] \rangle, \tag{5.79}$$

and

$$\Lambda(\vec{q}, \omega) = i \int_0^\infty dt e^{i\omega t} \langle [j(\vec{q}, t), j(-\vec{q}, 0)] \rangle = \chi_{jj}(\vec{q}, \omega), \tag{5.80}$$

is the longitudinal current–current susceptibility. From equation (5.29) we can write

$$\Lambda(\vec{q}, \omega) = -2\pi G^R(\vec{q}, \omega). \tag{5.81}$$

The dynamical conductivity $\sigma(q, \omega)$ is defined via

$$\langle j \rangle = \sigma(\vec{q}, \omega) iq_x h(\vec{q}, \omega). \tag{5.82}$$

We also define $\sigma(\omega) = \sigma(0, \omega)$. The formula for the conductivity in the long wavelength limit is therefore given by

$$\sigma(\omega) = \frac{1}{i\omega} (\langle K \rangle + \Lambda(\vec{q} = 0, \omega)). \tag{5.83}$$

The real part of σ can be written as

$$\sigma'(\omega) = D\delta(\omega) + \sigma^{\text{reg}}(\omega), \tag{5.84}$$

where the Drude weight D is given by

$$D = \pi\{\langle K\rangle + \Lambda'(\vec{q} = 0, \omega = 0)\}. \tag{5.85}$$

The regular part of the conductivity is

$$\sigma^{\text{reg}}(\omega) = \Lambda''(\vec{q} = 0, \omega)/\omega. \tag{5.86}$$

We can generalize the former procedure to fields which have a spatial dependence. For an external field $\phi_0(\vec{x}, t)$ coupled to an operator $O(\vec{x}, t)$, the interaction Hamiltonian is written as

$$H_{\text{int}} = -\int d\vec{x}\,\phi_0(\vec{x}, t)O(\vec{x}, t). \tag{5.87}$$

For simplicity we perturb with the operator O, and also measure the operator O. We have (see equation 5.27)

$$\delta\langle O\rangle = \int d^{d+1}x'\,G^R(x, x', t, t')\phi_0(x', t'), \tag{5.88}$$

where now

$$G^R(x, x', t, t') = i\theta(t - t')\langle[O(x, t), O(x', t')]\rangle, \tag{5.89}$$

is the retarded Green's function generalized from equation (5.25). Fourier transforming we find

$$\langle O(\omega, \vec{k})\rangle = -G^R(\omega, \vec{k})\phi(\omega, \vec{k}). \tag{5.90}$$

We define a transport coefficient χ by

$$\chi = -\lim_{\omega\to 0}\lim_{k\to 0}\frac{1}{\omega}G^R(\omega, \vec{k}). \tag{5.91}$$

In the case where $O = j^\mu$ is a conserved current, $\phi_0 = A_\mu$ is the boundary limit of a bulk gauge field, and the transport coefficient is the conductivity. If ϕ_0 in equation (5.88) is an external vector potential, let us say A_x, O is the conserved current J_x. Ohm's law for an electric field that is constant in space but oscillating in time with frequency ω, can be expressed as $J(\omega) = \sigma(\omega)E(\omega)$, where $\sigma(\omega)$ is the conductivity. The electric field is given by $E_x = -\partial_t A_x$. If $A_x \approx e^{-i\omega t}$, we have $E_x = i\omega A_x$. Using equation (5.90) we obtain

$$\sigma(\omega) = \frac{G^R(\omega, k = 0)}{i\omega}. \tag{5.92}$$

In classical mechanics the derivative of an on-shell action with respect to the boundary value of a field is simply equal to the canonical momentum conjugate to the field, evaluated at the boundary. This suggests the following definition:

$$\langle O \rangle = \frac{\delta S\left[\phi_0\right]}{\delta \phi_0} \equiv \lim_{r \to 0} \Pi\left[\phi_0\right]. \tag{5.93}$$

To obtain a finite action we need to add boundary terms to the action (see [6] and section 3.1). We can think of Π as the bulk field momentum, with r thought of as time.

$$G_{OO}^{R} = \lim_{r \to 0} \frac{\delta \Pi}{\delta \phi}\bigg|_{\delta \varphi = 0}. \tag{5.94}$$

For a relativistic quantum critical system with $\mu = 0$, the retarded Green's function

$$G_{\mu\nu}^{R}(x - x') = i\theta(x^0 - y_0)\left\langle\left[J_\mu(x), J_\nu(y)\right]\right\rangle, \tag{5.95}$$

at zero temperature, where here x and y are the space and time coordinates, is constrained by rotational invariance and charge conservation to be proportional to the projector onto conserved vectors:

$$P_{\mu\nu} = \eta_{\mu\nu} - \frac{K_\mu K_\nu}{K^2}, \tag{5.96}$$

where $K^2 = -K_0^2 + K^2$. All components of the Green's function are thus determined by a single scalar function

$$G_{\mu\nu}^{R}(K) = P_{\mu\nu}\Pi(K^2). \tag{5.97}$$

At finite temperature, the Lorentz invariance is broken and $P_{\mu\nu}$ can be split into transverse and longitudinal parts: $P_{\mu\nu} = P_{\mu\nu}^{L} + P_{\mu\nu}^{T}$, where

$$P_{\mu\nu}^{L} = \eta_{\mu\eta} - \frac{P_\mu P_\nu}{K^2} - P_{\mu\nu}^{T} \qquad P_{ij}^{T} = \delta_{ij} - \frac{K_i K_j}{K^2} \quad \text{and} \quad P_{t\mu}^{T} = 0. \tag{5.98}$$

References

[1] Forster D 1975 *Hydrodynamics Fluctuations, Broken Symmetry and Correlation Functions* (New York: Benjamin)
[2] Hartnoll S A Quantum critical dynamics from black holes arXiv:0909.3553 [cond-mat.str-el]
[3] Sachdev S 1999 *Quantum Phase Transitions* (Cambridge: Cambridge University Press)
[4] Sentef M, Kolar M and Kampf A P 2007 Spin transport in Heisenberg antiferromagnets in two and three dimensions *Phys. Rev.* B **75** 214403

Pires A S T and Lima L S 2009 Spin transport in antiferromagnets in one and two dimensions calculated using the Kubo formula *Phys. Rev.* B **79** 064401

[5] Scalapino D J, White S R and Zhang S 1993 Insulator, metal, or superconductor: the criteria *Phys. Rev.* B **47** 7995

[6] Hartnoll S A 2009 Lectures on holographic methods for condensed matter physics *Class. Quantum Grav.* **26** 224002

Chapter 6

Dynamics using the AdS/CFT formalism

6.1 Calculation of transport coefficients

Here we will discuss the bulk calculation of transport coefficients following references [1] and [2]. To obtain the retarded Green's function at finite temperature, one solves the bulk equations of motion for the field ϕ dual to O, and linearizes the equations about the black hole metric. Iqbal and Liu [2] consider a fictitious membrane at each constant-radius hypersurface (starting at the black hole) and introduce a linear response function for each of them. They then derive a flow equation for the radius dependent response function, and calculate how it evolves from the horizon to the boundary, where it determines the response of the dual field theory. To do that they start with the metric (4.38) written in a general form as

$$ds^2 = -g_{tt}dt^2 + g_{rr}dr^2 + g_{ij}dx^i dx^j. \qquad (6.1)$$

Near the horizon $r = r_+$ the metric may be written

$$g_{tt} = A(r - r_+) \qquad g_{rr} = \frac{B}{r - r_+}. \qquad (6.2)$$

For our case $A = L^2 d/r_+^3$, $B = L^2/r_+ d$, but for the moment I will keep the general result (6.1). The equation of motion (4.10), with $m = 0$, for this metric is given by

$$\frac{\partial}{\partial r}\left(\frac{r - r_+}{B}\frac{\partial \phi}{\partial r}\right) - \frac{1}{A(r - r_+)}\frac{\partial^2 \phi}{\partial t_2} = 0. \qquad (6.3)$$

doi:10.1088/978-1-627-05309-9ch6

Taking $\phi(r, t) = \phi(r)e^{i\omega t}$ leads to

$$\frac{A}{B}(r - r_+)\frac{\partial}{\partial r}(r - r_+)\frac{\partial \phi(r)}{\partial r} + \omega^2\phi(r) = 0. \tag{6.4}$$

Introducing a variable x such that

$$\frac{dx}{dr} = \sqrt{\frac{g_{rr}}{g_{tt}}}, \tag{6.5}$$

and using

$$\frac{g_{rr}}{g_{tt}} = \frac{B}{(r - r_+)^2 A}, \tag{6.6}$$

we find

$$\frac{\partial^2\phi(x)}{\partial x^2} + \omega^2\phi(x) = 0. \tag{6.7}$$

The solution is $\phi(x, t) \propto e^{-i\omega(t\pm x)}$.

The in-falling boundary condition implies that we should take the positive sign in the exponent. Using a coordinate v defined by $v = x + t$, and integrating the equation

$$dv = dt + \sqrt{\frac{g_{rr}}{g_{tt}}}\,dr, \tag{6.8}$$

we obtain

$$v = t + \frac{1}{4\pi T}\ln(r - r_+), \tag{6.9}$$

where I have used

$$\frac{1}{T} = 4\pi\sqrt{\frac{A}{B}}. \tag{6.10}$$

The final result is

$$\phi(r, t) \propto (r - r_+)\exp\left(-\frac{i\omega}{4\pi T}\right)e^{-i\omega t}. \tag{6.11}$$

In the coordinates (v, r) the metric is written as

$$ds^2 = A(r - r_+)dv^2 - 2\sqrt{AB}\,dv dr. \tag{6.12}$$

In these coordinates the metric is non-singular at $r = r_+$.

To proceed we start with the action at the horizon

$$S_H = -\int_\Sigma d^d x \frac{\sqrt{g}}{q(r)}g^{rr}(\partial_r\phi)\phi(r_+, x), \tag{6.13}$$

where $q(r)$ is an effective scalar coupling. In practice, Σ is a time-like surface of fixed r just outside the true horizon. Multiplying and dividing by $\sqrt{\gamma}$ where $\gamma_{\mu\nu}$ is the induced metric on the horizon Σ we have

$$S_{\mathrm{H}} = \int_{\Sigma} d^d x \sqrt{\gamma} \left[-\frac{\sqrt{g}}{\sqrt{\gamma}\, q(r)} g^{rr} \partial_r \phi \right] \phi(r_+, x). \tag{6.14}$$

Using equation (5.93) we see that the momentum conjugate to ϕ (r direction) is

$$\Pi = -\frac{\sqrt{g}}{q(r)} g_{rr} \partial_r \phi. \tag{6.15}$$

The horizon is a regular place for in-falling observers. For these observers ϕ is not singular. This means that ϕ can depend on r and t only through the coordinate v, that is $\phi(r, t, x) = \phi(v, x_i)$. We have

$$\frac{d}{dr} \rightarrow \sqrt{\frac{g_{rr}}{g_{tt}}} \partial_t \phi, \tag{6.16}$$

leading to

$$\Pi = -\frac{\sqrt{g}}{q(r)} g^{rr} \partial_t \phi. \tag{6.17}$$

Using $g^{rr} = 1/g_{rr}$, and $\phi \propto e^{-i\omega t}$, we arrive at the final result

$$\Pi(r, k_\mu) = \frac{1}{q(r)} \sqrt{\frac{g}{g_{rr} g_{tt}}} i\omega \phi(r_+, k_\mu). \tag{6.18}$$

It can be shown [2] that in the limit $k_\mu \rightarrow 0$, $\omega \rightarrow 0$, we have

$$\left. \frac{\Pi}{\omega\phi} \right|_{r=0} = \left. \frac{\Pi}{\omega\phi} \right|_{r=r_+}. \tag{6.19}$$

Thus

$$\chi = \frac{1}{q(r_+)} \sqrt{\frac{g}{g_{rr} g_{tt}}} \bigg|_{r_+}, \tag{6.20}$$

where

$$g = g_{tt} g_{rr} \left(\frac{1}{r_+^2} \right)^{d-1}. \tag{6.21}$$

In the case where $O = T_y^x$, the source is the boundary value for a metric perturbation δg_x^y, and the transport coefficient χ is the shear viscosity η. Viscosity is associated with the tendency of a substance to resist flow. A perfect fluid has negligible shear viscosity. Lower shear viscosities are associated with strongly interacting particles. As we have seen, the entropy density s of the boundary field theory is given by $s = A/4G_N V_{d-1}$, which leads to

$$\frac{\eta}{s} = \frac{4G_N}{q(r_+)}. \tag{6.22}$$

In the case of Einstein gravity $q = 16\pi G_N$, we get

$$\frac{\eta}{s} = \frac{1}{4\pi}. \tag{6.23}$$

The ratio η/s gives a measure of the interaction per constituent that better allows a comparison across different systems at widely different scales. The horizon response always corresponds to that of the low frequency limit of the boundary theory (regardless of the specific model we use). Away from this limit, the full geometry of the space time becomes important. The relation (6.23) provides a lower bound in a large class of systems. The predicted value of this ratio for the quark–gluon plasma was confirmed at the Relativistic Heavy Ion Collider at Brookhaven National Laboratory in 2008 [2].

Iqbal and Liu [2] have also derived expressions for various transport coefficients in terms of the components of the metric evaluated at the horizon. These coefficients are determined by universal constants of nature, and not by collision rate.

6.2 Dynamics close to equilibrium

In this section, I will study the heat current in the context of AdS/CFT correspondence. For simplicity I will take $d = 3$, a zero chemical potential and consider the limit of zero momentum. The heat current (which is the response of the system to a temperature gradient) can be written as

$$\langle Q_x \rangle = \langle T_{tx} \rangle = -k\frac{\partial T}{\partial x}, \tag{6.24}$$

where k is the thermal conductivity and the current is taken pointing in the x direction.

The period of Euclidean time is $1/T$, and therefore we can rescale the time so that there is no temperature dependence in the period. Writing $t = t/T$, the metric, in Minkowski space and in this coordinate, becomes $g_{tt} = -1/T^2$. A small thermal gradient $T \to T + x\partial_x T$ implies

$$\delta g_{tt(0)} = -\frac{2x\partial_x T}{T^3}. \tag{6.25}$$

Going back to the original time t, we have

$$\frac{\partial g_{tt(0)}}{\partial x} \approx -\frac{2\partial_x T}{T}. \tag{6.26}$$

Now taking g_{tt} as a constant we perform a change of coordinates in which the temperature gradient is exhibited by a fluctuation of an off-diagonal component of the metric (the energy transport is related to the T_{tx} component of the stress-energy tensor). Under an infinitesimal coordinate transformation $x^\alpha \to x^\alpha + \xi^\alpha$, the metric changes by

$$\delta g_{\mu\nu} = \partial_\mu \xi_\nu + \partial_\nu \xi_\mu. \qquad (6.27)$$

This can be verified using the transformation law for the metric components

$$g_{\hat\mu\hat\nu} = \frac{\partial x^\mu}{\partial x^{\hat\mu}} \frac{\partial x^\nu}{\partial x^{\hat\nu}} g_{\mu\nu}. \qquad (6.28)$$

Following Herzog [3] we can choose ξ_μ such that $\partial_x (g_{tt} + \delta g_{tt}) = 0$. Setting $\xi_x = 0$, and using $\delta g_{tx} = \partial_t \xi_x + \partial_x \xi_t = \partial_x \xi_t$, we obtain $\partial_x g_{tt} + \partial_x \delta g_{tt} = 0$. But $\partial_x \delta g_{tt} = \partial_x(\partial_t \xi_t + \partial_t \xi_t) = 2\partial_t \partial_x \xi_t$, leading to $2\partial_t \partial_x \xi_t = -\partial_x g_{tt}$. If we take a time dependence of the form $e^{-i\omega t}$ we obtain

$$\partial_x \xi_t = \frac{\partial_x g_{tt(0)}}{2i\omega} = -\frac{\partial_x T}{i\omega T}, \qquad (6.29)$$

which gives

$$\delta g_{tx(0)} = -\frac{\partial_x T}{i\omega T}. \qquad (6.30)$$

From equations (6.24) and (6.30) we can write

$$\langle Q_x \rangle = kTi\omega \delta g_{tx(0)}. \qquad (6.31)$$

The next step is to solve Einstein's equation of motion for the perturbation δg_{tx} in the bulk. Linearizing this equation about the background solution for the four-dimensional black hole, we find

$$\frac{dh}{dr} + \frac{2h}{r} = 0, \qquad (6.32)$$

where $h = \delta g_{tx}$. The solution is $h = ar^{-2}$, where a is a constant. Using the action (4.62) we obtain

$$\Pi_{g_{tx}} = \frac{\delta S}{\delta g_{tx(0)}} = \frac{L^2}{4\pi G_N r^3}\left(1 - \frac{1}{\sqrt{f}}\right)\delta g_{tx(0)}, \qquad (6.33)$$

where f is given by equation (4.49), and $g_{tx(0)}$ is defined at finite r by $g_{tx(0)} = (r/L)^2 g_{tx}$. Using equation (5.93) we obtain in the limit $r \to 0$

$$\langle T_{tx} \rangle = -\varepsilon \delta g_{tx(0)}, \qquad (6.34)$$

where $\varepsilon = L^2/8\pi G_N r_+^3$. Comparing (6.31) with (6.34) gives

$$k(\omega) = \frac{i\varepsilon}{\omega T}. \tag{6.35}$$

The case of a finite chemical potential was studied by Hartnoll [1].

6.3 Spin transport

In this section I will briefly discuss spin transport in the model presented in section 2.2. More details can be found in references [4] and [5]. Spin transport is a very interesting problem in condensed matter physics. In an insulating magnet, magnetization can be transported by excitations such as magnons and excitons without the transport of charge. As it is expected that transport properties in the critical region should have a universal character, we can use a CFT, which is solvable by AdS/CFT correspondence, to obtain results that can be used in our model.

While charge conductivity is studied as the current response to a time dependent electromagnetic potential, the spin current flows in response to a magnetic field gradient, which plays the role of the chemical potential for spins. Spin transport for the model described by the Hamiltonian (2.7) was studied for $D > D_C$ and $D < D_C$ in references [5] and [6] respectively. In the critical region, spin transport can be studied using an expression similar to equation (2.9), including the effect of the magnetic field gradient [4].

The spin conductivity, $\sigma(\omega)$, is related to the retarded correlation function, $\chi(k, \omega)$, by the Kubo formula

$$\sigma(\omega) = -\lim_{k \to 0} \frac{i\omega}{k^2} \chi(k, \omega). \tag{6.36}$$

It can be shown that at high frequencies or low temperatures ($\omega/T \gg 1$), $\chi(k, \omega)$, in all CFTs in spatial dimensions higher than one, reduces to the $T = 0$ limit, where scale invariance, Lorentz symmetry and 'charge' conservation determine $\chi(k, \omega)$ up to a constant [7, 8]:

$$\chi(k, \omega) = \text{const.} \frac{Kk^2}{\sqrt{v^2 k^2 + (\omega + i\eta)^2}} \qquad \hbar\omega \gg k_B T, \tag{6.37}$$

where η is the viscosity, K is a universal number, and v is the speed of the excitations.

In the low frequency limit ($\hbar\omega \ll k_B T$) one expects a diffusive behavior, and we have

$$\chi(k, \omega) = Q^2 \frac{\chi_C D k^2}{Dk^2 - i\omega}, \tag{6.38}$$

where χ_C is the compressibility, D is the diffusion constant and for spin transport the 'charge' Q is given by [6] $Q = g\mu_B/4\sqrt{\hbar}$. Quantum critical scaling arguments show that [7, 8]

$$\chi_C = \Theta_1 k_B T/(h\nu)^2 \qquad D = \Theta_2 h\nu^2/(k_B T), \qquad (6.39)$$

where Θ_1 and Θ_2 are universal numbers.

The first exact results for a quantum critical point in $(2+1)$ dimensions were obtained in supersymmetric Yang–Mills non-Abelian gauge theory, using AdS/CFT correspondence [9]. Exact results for the full structure of $\chi(k, \omega)$ were obtained and the expected limiting forms in equations (6.37) and (6.38) were obeyed. Exact results for K, Θ_1 and Θ_2 were found. A curious feature in this theory was the relation $K = \Theta_1\Theta_2$.

References

[1] Hartnoll S A 2009 Lectures on holographic methods for condensed matter physics *Class. Quantum Grav.* **26** 224002
[2] Iqbal N and Liu H 2009 Universality of the hydrodynamic limit in AdS/CFT and the membrane paradigm *Phys. Rev.* D **79** 025023
[3] Herzog C P 2009 Lectures on holographic superfluidity and superconductivity *J. Phys. A: Math. Gen.* **42** 343001
[4] Sachdev S 1999 *Quantum Phase Transitions* (Cambridge: Cambridge University Press)
[5] Lima L S and Pires A S T 2009 Dynamics of the anisotropic two-dimensional XY model *Eur. Phys. J.* B **70** 335
[6] Pires A S T and Lima L S 2010 Spin transport in the anisotropic easy-plane two-dimensional Heisenberg antiferromagnet *J. Mag. Mag. Mat.* **322** 668
[7] Herzog C P, Kovtun P, Sachdev S and Son D T 2007 Quantum critical transport, duality, and M-Theory *Phys. Rev.* D **75** 085020
[8] Sachdev S Finite temperature dissipation and transport near quantum critical points arXiv:0910.1139 [cond-mat]
2011 The landscape of the Hubbard model *TASI 2010 String Theory and its Applications: From meV to Planck Scale*, ed M Dine, T Banks and S Sachdev (Singapore: World Scientific) arXiv:1012.0299 [hep-th]
[9] Itzhaki N, Maldacena J M, Sonnenschein J and Yankielowicz S 1998 Supergravity and the large N limit of theories with sixteen supercharges *Phys. Rev.* D **58** 046004

Chapter 7

Bosons and fermions

7.1 Holographic superconductors

Let us remind ourselves of what we have done up to now. In section 4.2 we studied a scalar field in empty space. In section 4.4, we considered electromagnetic (EM) fields in the presence of a black hole. In this section we will go one step further and analyze a scalar complex field together with EM fields in the presence of a black hole. Several interesting questions will be raised. This model is of interest because it describes what is called a holographic superconductor. Thus I will start this section by giving a brief overview of superconductivity.

A lot of interest in condensed matter physics is dedicated to the study of ordered phases at low temperatures. These phases appear as a consequence of the formation of a symmetry breaking condensate. Therefore it is of interest to study how the AdS/CFT formalism induces symmetry breaking phase transitions. We remark that these are not QPTs but ordinary thermal phase transitions.

One of the most interesting examples of one of these transitions is the emergence of superconductivity. Conventional superconductors are well described by BCS theory [1], where electrons bind (due to interactions with photons) to form Cooper pairs (which are charged bosons), with opposite spins. Below the critical temperature T_C there is a second order phase transition and these bosons condense, leading to an infinite DC conductivity. In this phase there is also an energy gap for charged excitations. In BCS theory the electron–phonon interaction is weak, and Cooper pairs are much larger than the inter-atomic spacing.

In 1986 a new class of high T_C superconductors was discovered in cuprates, where the superconductivity is along the CuO_2 planes [2]. In 2008 another class of super-conductors was discovered based on iron, the so called iron pnictides [3]. These materials are also layered and superconductivity is again in two-dimensional planes. It is believed that electron pairs are responsible for the high T_C, but coupling is not weak and the mechanism, up to now, is not well understood. Unlike BCS theory, this mechanism involves strong coupling.

One motivation for building a holographic superconductor is to have a microscopic description of the onset of superconductivity in which there are no quasi-particles. Instead, there is a strongly coupled theory in which a charged operator condenses below a critical temperature. Instead of quantities such as 'electrons' and 'pairing', the onset of superconductivity in holographic models is mediated by charge and scaling dimensions of operators in quantum critical theory [4]. As in BCS theory, fermion pairs are formed at T_C, and a gap appears below T_C; however, with BCS theory the normal state is a well-developed Fermi liquid. In a holographic superconductor fermion pairs are formed at T_C, however a pairing instability arises in a non-Fermi liquid where the Cooper mechanism is not operative. If the theory is not correct, at least it may give an understanding of some basic features of high temperature superconductors. The holographic mechanism (implemented mainly by Hartnoll, Herzog and Horowitz) is viewed as the first mathematical theory for the mechanism of superconductivity that goes beyond BCS theory and the holographic superconductor is considered to the first success of AdS/CFT at addressing the physics of finite density matter. To build a holographic superconductor the first step is to find the gravitational dual for a superconductor [5–24].

Spontaneous symmetry breaking occurs if a charged operator in the boundary acquires vacuum expectation values. Equation (4.5) implies that such a charged operator will be in duality with charged fields in the bulk [25]. A charged AdS black hole corresponds to the normal high temperature phase of a holographic superconductor. However, at low temperatures instability will occur and the Reissner–Nordström black hole is no longer the correct ground state. One must therefore solve the coupled Einstein–Maxwell scalar equations of motion to find new solutions in which the scalar field is different from zero. There are theorems demonstrating that in asymptotically flat space time the Einstein equations allow only for black holes that are characterized by mass, charge and angular momentum. That is, we cannot have a scalar field associated with the black hole. John Wheeler invented the phrase *A black hole has no hair* to describe these theorems. (Thus, a static non-zero field outside a black hole is called black hole hair.) However, the no-hair theorems do not apply to asymptotically AdS space time; in this case, we say that the black hole has scalar hair at low temperature, but no hair at high temperatures.

If the electric field at the horizon of the charged black hole is large enough, pairs of positive and negative charged particles (quanta for the field ψ introduced below) can be created near the horizon. The particles with opposite charge to the black hole fall into the horizon, reducing the charge of the black hole. The particles with the same sign charge as the black hole are repelled from it. In asymptotically flat space time, these particles escape to infinity, so the final result is the Reissner–Nordström black hole. In AdS space, the charged particles cannot escape since the negative cosmological constant acts like a confining box, and they settle hovering outside the horizon. This gas of charged particles is the quantum description of hair [26].

To achieve the superconducting phase, we need to include in the Einstein–Maxwell action additional complex scalar charged fields and we arrive at the model mentioned in the beginning of this section. To simplify, it is usual to consider only a single charged field in the bulk, coupled to a charged scalar operator. For more general superconductors the reader can consult reference [10].

The minimal Lagrangian (in $(d+1)$ dimensions) describing an AdS charged black hole and a scalar field is given by

$$L = \frac{1}{2k^2}\left[R + \frac{d(d-1)}{L^2}\right] - \frac{1}{4e^2}F^{\mu\nu}F_{\mu\nu} - |\nabla\psi - iqA\psi|^2 - m^2|\psi|^2, \qquad (7.1)$$

where k^2 is defined after equation (4.68), q is the charge of the scalar field and $e = g_4$ is the Maxwell coupling constant. The action is given by

$$S = \int d^d x \sqrt{g}\, L. \qquad (7.2)$$

We will consider the case of $d = (2+1)$ dimensions. In equation (7.1) we considered a simplification of a more general potential:

$$V(\psi) = m^2\psi\psi^* + \frac{\lambda(\psi\psi^*)^2}{2} + \cdots \qquad (7.3)$$

The nature of the solutions depends significantly on the potential $V(\psi)$. If ψ falls off sufficiently quickly (i.e. goes to zero or a constant) near the boundary, the back reaction on the metric becomes negligible and the space time is asymptotically AdS.

As is well known, there are no classical phase transitions in two-spatial-dimension field theories (except the Kosterlitz–Thouless transition). The holographic models get around this obstacle by having a large number N degrees of freedom [5].

We know that the scalar field ϕ in the bulk is in duality with an operator O in the boundary. (The symmetry breaking field in the boundary is a pair field of fermions, as in BCS theory.)

The vacuum expectation value for this operator should stay non-zero even when the external field breaking the symmetry is turned off. That is, a charged condensate corresponds to a non-vanishing expectation value $\langle O \rangle$ for the charged operator in duality with the bulk field ψ. If the charge q is zero, equation (7.1) gives a theory for free scalar particles, free electromagnetic (EM) fields interacting weakly (hence the term free) and gravity. Each of these terms has been treated before separately. If q is different from zero there are interactions between the charged scalar field and the EM field, and the equations of motion for varying S are non-linear coupled differential equations involving ϕ, A_μ and the background metric $g_{\mu\nu}$. These equations are [27]:

$$R_{\mu\eta} - \frac{1}{2}g_{\mu\eta}R - \frac{3}{L^2}g_{\mu\eta} = k^2\left[\frac{1}{e^2}\left(F_{\mu\lambda}F_\mu{}^\lambda - \frac{1}{4}F^2 g_{\mu\eta}\right) + (\partial_\mu + iqA_\mu)\psi^*(\partial_\eta - iqA_\eta)\psi\right]$$

$$+ k^2\left[(\partial_\mu - iqA_\mu)\psi(\partial_\nu \pm qA_\nu)\psi^* - g_{\mu\nu}(|(\partial_\lambda - iqA_\lambda)\psi|^2 + m^2|\psi|^2\right]. \qquad (7.4)$$

The equation of motion for the field ψ is

$$\left(\nabla^\nu - iqA^\nu\right)\left(\nabla_\nu - iqA_\nu\right)\psi - m^2\psi = 0. \tag{7.5}$$

And the Maxwell equation is

$$\nabla_\nu F^{\mu\nu} = iqe^2\left[\psi^*\left(\partial^\nu - iqA^\nu\right)\psi - \psi\left(\partial^\nu + iqA^\nu\right)\psi^*\right]. \tag{7.6}$$

The term in the Lagrangian (7.1)

$$m_{\text{eff}}^2 = m^2 + g^{tt}q^2\psi^2, \tag{7.7}$$

(where I have used $|A|^2 = g^{\mu\nu}A_\mu A_\nu$) can be interpreted as an effective mass squared for the scalar field ψ. If m is small the gauge field term may dominate, and since g^{tt} is negative the effective mass becomes tachyonic. Combining equation (7.7) with the Breitenlohner–Freedman condition $m^2L^2 \geqslant -d^2/4$ (discussed in section 4.2), Hartnoll [25] has shown that there is a range of parameters for the bulk field ψ in which the normal phase becomes unstable as the temperature is lowered, corresponding to the holographic superconductors. A theory with gapless charged bosonic excitations will develop condensates at weak coupling. At strong coupling this may not happen (if there are no operators in the range found by Hartnoll) and it is possible that these theories can provide realizations of exotic *Boson metal phases* [4]. Bose condensation (different from flat space time) is driven by a change of sign in m^2.

We look now for solutions with an asymptotic AdS geometry at the boundary. First we consider the normal high temperature phase. This, as we know, is the dual of a solution to the equation of motion derived from (7.1) with $\psi = 0$. We remember that the metric (4.48) corresponding to a scale invariant theory at finite temperature does not give a preferred critical temperature T_C at which something different happens. All temperatures are equivalent. As proposed by Hartnoll *et al* [6], the simplest way to introduce a scale is to work at a finite chemical potential μ, such that $T_C \propto \mu$ and therefore, as we saw in section 4.4, the theory is the dual of the Reissner–Nordström black hole.

In the action (7.2) the fields contribute to the curvature of the space time and this effect, called the back reaction, has to be taken into account. Rescaling $\psi \to \psi/q$, $A \to A/q$, the matter action acquires a $1/q^2$ term in front, so that large q suppresses the back reaction on the metric. We say that we are dealing with a probe and the metric can be taken as the one given by equation (4.48).

We assume we have only an electric field in the bulk, i.e. $A_t = \phi(r)$, $A_r = A_x = A_y = 0$, and $\psi = \psi(r)$. The Maxwell equation implies that the phase of ψ must be constant, so we can assume ψ is real. The field equations then take the form

$$r^4\left(\frac{d^2\psi}{dr^2} + \frac{1}{f}\frac{df}{dr}\frac{d\psi}{dr}\right) + \frac{\phi^2}{f^2}\psi - \frac{m^2}{2L^2}\psi = 0. \tag{7.8}$$

Imposing the Lorentz gauge condition $\nabla_t A^t = 0$ the equation for ϕ is given by

$$r^4 \frac{d^2 \phi}{dr^2} - \frac{2\psi^2}{f} \phi = 0. \tag{7.9}$$

We insert the function $f(r)$ given by (4.76) into (7.8) and (7.9) and obtain two coupled singular differential equations for ψ and ϕ. These equations have diverging terms as r approaches the horizon and the boundary. In order to solve these equations we put boundary conditions on the horizon and integrate out to the boundary where we read off the solutions (see equation (7.10) below). We are interested in solutions that are regular at the horizon and normalizable at the boundary. To get around the problem of singularities, we can use the Frobenius method. This allows us to construct a series expansion around the singular points and determine order by order the coefficients in the expansion. The field ϕ has to be zero on the horizon to avoid an infinite norm at this point. After that we solve the differential equation using a numerical integrator (such as the routine NDsolve in Mathematica) in the rest of the bulk. For more details see the appendix of reference [24] where the method is very well explained. In figure 7.1 we show a sketch of the bulk solution along the radial direction $u \propto 1/r$ (measured from the black hole). Near the boundary the solutions behave as

$$\psi(r) = r\psi^{(1)} + r^2\psi^{(2)},$$
$$\phi(r) = \mu - \rho r, \tag{7.10}$$

where μ is the chemical potential and ρ the charge density. The overal signs of ψ and ϕ are not fixed. Spontaneous symmetry breaking means that there is a vacuum expectation value for an operator, which stays non-zero even when the external field breaking the symmetry is turned off. We impose the condition that either $\psi^{(1)}$ or $\psi^{(2)}$ vanishes. (The condition that both $\psi^{(1)}$ and $\psi^{(2)}$ are non-zero leads to a theory in which the asymptotic AdS region is unstable [6].) We can interpret $\psi^{(1)}$ as a source for the operator in the field theory, while $\psi^{(2)} \propto \langle O_2 \rangle$ is an expectation value.

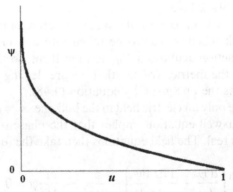

Figure 7.1. The scalar hair along the radial direction u (in arbitrary units). Adapted from [27].

We are interested in solutions with no source for ψ, i.e. $\psi^{(1)} = 0$, but where $\psi^{(2)}$ becomes non-zero at some critical temperature. Otherwise, we take $\psi^{(1)} \propto \langle O_1 \rangle$ and set $\psi^{(2)} = 0$. The Hawkings temperature can be calculated as before and is given by $T = 1/4\pi\sqrt{\rho}$.

The condensate of the scalar operator O in the field theory in duality with the field ψ is given by

$$\langle O_i \rangle = \sqrt{2}\psi^{(i)} \qquad i = 1, 2. \tag{7.11}$$

The factor $\sqrt{2}$ was introduced in [6] for convenience. It is easy to check that $\langle O_1 \rangle / T$ and $\sqrt{\langle O_2 \rangle}/T$ are dimensionless quantities. Equations (7.8) and (7.9) have been solved numerically in [6]. (Note that they used $r \to 1/r$.) Near the transition, they found a square root behavior (typical of second order phase transitions):

$$\begin{aligned}
\langle O_1 \rangle &\approx 9.3 T_C (1 - T/T_C)^{1/2}, &\text{as} \quad T \to T_C &\approx 0.226\rho^{1/2} \\
\langle O_2 \rangle &\approx 144 T_C^2 (1 - T/T_C)^{1/2}, &\text{as} \quad T \to T_C &\approx 0.118\rho^{1/2}.
\end{aligned} \tag{7.12}$$

Computing the free energy (Euclidean action) of the hairy configuration and comparing with the solution $\psi = 0$, $\phi = \mu - \rho r$, the authors of [6] found that the free energy is always lower for the hairy configuration and becomes equal as $T \to T_C$. The plot for $\sqrt{\langle O_2 \rangle}/T_C$ as a function of T/T_C, presented in [6, 24], is very similar to the energy gap curve predicted using BCS theory. However the curve tends toward ∞ as $T \to 0$ and the BCS curve tends to 3.5. Also, the gap for the high T_C superconductors takes values between 6 and 10 (figure 7.2).

As with BCS theory the phase transition of the holographic superconductor is a mean field, but for the holographic superconductor this is caused by the large N limit. It is expected that a $1/N$ computation involving quantum corrections in the bulk will lead to corrections to the mean field exponents. The bulk solution is rotationally invariant and therefore it describes an s-wave superconductor. There is evidence that the iron based superconductors may be s-wave.

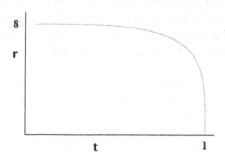

Figure 7.2. $r = \sqrt{\langle O_2 \rangle}/T_C$ as a function of $t = T/T_C$.

To calculate the conductivity we consider perturbations of the vector potential in the x direction (for instance) and assume a time dependence of the form $e^{-i\omega t}$. The equation for the A_x component is given by

$$\frac{d}{dr}\left(f\frac{dA_x}{dr}\right) - \frac{\omega^2}{f}A_x = \frac{2e^2}{r^2}\psi^2 A_x. \tag{7.13}$$

Close to the boundary the solution for A_x behaves as

$$A_x = A_x^{(0)} + rA_x^{(1)}. \tag{7.14}$$

The GKPW rule gives us

$$A_x = A_x^{(0)} \qquad \langle J_x \rangle = A_x^{(1)} \tag{7.15}$$

for the dual source and expectation values for the current. Ohm's law gives

$$\sigma(\omega) = \frac{\langle J_x \rangle}{E_x} = -\frac{\langle J_x \rangle}{(dA_x/dt)} = -i\frac{A_x^{(1)}}{\omega A_x^{(0)}}. \tag{7.16}$$

The conductivity σ as a function of ω/T was computed in [6] and it was found that the DC conductivity became infinite below T_C and the frequency dependent conductivity developed a gap determined by the condensate. Zaanen *et al* [27] have pointed out that the only way to distinguish sharply between the mechanisms for BCS and holographic superconducting in experiments is by measuring the dynamical pair susceptibility in the normal state.

In reference [7] Hartnoll *et al* took into account the effect of the back reaction on the metric. They started with the following metric ansatz for a hairy black hole, below the critical temperature at which ψ becomes unstable and non-vanishing:

$$ds^2 = \frac{L^2}{r^2}\left(-f(r)e^{-\chi(r)}dt^2 + \frac{dr^2}{f(r)} + dx^i dx^i\right), \tag{7.17}$$

together with $A = A_t(r)$, $\psi = \psi(r)$.

Inserting these expressions in the equation of motion (7.4)–(7.6), and solving the equations numerically they calculated $f(r)$, $\chi(r)$, A and ψ. The results are presented and discussed in [7], where the conductivity is also calculated. They found that the qualitative behavior was similar for all charges, but in the limit of arbitrarily small charges the effect of the back reaction produced some changes with respect to the probe limit discussed before. Holographic superconductors in $d = (3 + 1)$ dimensions were studied in [8].

The exclusion of magnetic fields in a superconductor, when it is cooled below its critical temperature, is called the *Meissner* effect. Materials in which this effect is complete are known as *Type I superconductors*. In *Type II superconductors*, the superconductivity is observed up to an upper critical field H_{c2}, but an incomplete Meissner effect is observed between a lower critical field H_{c1} and H_{c2} [28]. All high T_C cuprate superconductors are Type II. Holographic superconductors are also Type II.

At zero temperature the hairy black hole disappears and what remains is a 'lump' of scalar field that has absorbed the black hole charge. Zaanen *et al* [27] have called this object a 'Higgs lump'. The object has zero horizon area (and therefore zero entropy) consistent with a non-degenerate ground state. With increasing m, while fixing other parameters, a QPT occurs at zero temperature from the holographic superconductor to the Reissner–Nordström zero temperature metal [27].

If the charged matter is bosonic we obtain a holographic superconductor. On the other hand, fermions cannot macroscopically occupy their ground state as a consequence of the Pauli Exclusion Principle. As is well known from solid state physics, a large Maxwell potential A_t does not cause Bose–Einstein condensation, but rather leads to the formation of a Fermi surface. The filled Fermi sea is a quantum vacuum of the fermions in the presence of a chemical potential. Neutron stars can be considered as gravitating Fermi surfaces, and for this reason the solutions for holographic charged fermion systems are called *electron stars*. For more details see [29].

7.2 Fermions and AdS/CFT correspondence

Let us start with a gas of non-interacting spin ½ fermions containing N particles in a box. Fermions are characterized by the Pauli Exclusion Principle and so the ground state is obtained by filling all the single-particle states inside a sphere with radius k_F in momentum space. The boundary of this sphere is called the *Fermi surface*. The low energy excitations are given by either filling a state slightly outside the Fermi surface or removing a fermion from a filled state slightly inside the Fermi surface, and therefore creating a hole. The excitations are called *particles* and *holes* respectively. These excitations are gapless with an energy given by

$$\varepsilon_k = E_0(k) - E_F, \tag{7.18}$$

where $E_0(k) = k^2/2m$ is the energy of a free fermion particle with mass m, and $E_F = k_F^2/2m$. The chemical potential is given by $\mu = E_F$. We can also define a Fermi velocity by $v_F = k_F/m$. Equation (7.18) can thus be written as

$$\varepsilon_k = v_F(k - k_F). \tag{7.19}$$

The retarded Green's function for a gas of non-interacting fermions is given by

$$G^R(k, \omega) = \frac{1}{\omega - \varepsilon_k + i0_+}, \tag{7.20}$$

with Fourier transform

$$G^R(k, t) = \theta(t)e^{-i\varepsilon_k t}. \tag{7.21}$$

The case of interacting fermions is treated using Landau–Fermi liquid theory. This essentially states that [28]:
- The ground state of an interacting fermionic system is characterized by a Fermi surface in momentum space.

- The fundamental fermions may interact strongly, but the low energy excitations near the Fermi surface behave like weak particles and holes (called quasi-particles).

The dominant effect of electron interactions in a metal is to renormalize the effective mass of the electron. That is, the low energy degrees of freedom for a metal are like electrons, in that they are long lived fermionic particles carrying electric charge, but they are 'dressed' to account for the interactions between them. The energy of a quasi-particle is similar to that in free theory (7.19), but now $v_F = k_F/m^*$, where m^* is the effective mass. This effective mass can be very large. In heavy fermion metals it has a value of the order of hundreds of times the bare electron mass.

This theory is very useful because it can describe almost all the known metals. It is interesting that the theory works, despite the fact that the Coulomb interaction between electrons in ordinary metals is as large as the Fermi energy E_F. The theory implies that a perturbative expansion works. We can start with a free electron system and use perturbation theory to calculate various quantities. This is what is done in almost all books dealing with many body interactions, where the calculations are performed using Feynman diagrams.

A quasi-particle (or hole) can decay into another particle plus a number of particle–hole pairs; but near the Fermi surface it has a long lifetime [30]. Here we can write

$$G^R(k, t) \approx \theta(t) e^{-i\varepsilon_k t - \frac{\Gamma}{2} t}, \tag{7.22}$$

where $\Gamma \propto \varepsilon_k^2$. From (7.22) we obtain

$$G^R(k, \omega) \approx \frac{Z}{\omega - v_F(k - k_F) + \Sigma(k, \omega)}. \tag{7.23}$$

The term $\Sigma(k, \omega) \approx -i\Gamma/2$ is called the self-energy. From the spectral representation of the Green's function studied in section 5.5, we find that the residue Z of the pole is given by the overlap between the one-quasi-particle state with the state created when we apply the fermion creation operator on the vacuum. Near the Fermi surface, the spectral function

$$A(k, \omega) = \frac{1}{\pi} \mathfrak{I} \, G^R(k, \omega) \tag{7.24}$$

is given by a Lorentzian peak centered at ε_k with a width proportional to ε_k^2. For $k \to k_F$ it becomes a delta function.

In a metal there is a repulsive Coulomb interaction between the electrons, together with an attractive interaction between electrons and ions. When an electron moves, it tends to pull the ion towards it; when the ion moves in the direction of the electron, the other electrons tend to follow the ion and therefore the first electron (the vibration of the lattice is described in terms of phonons). This leads to an effective attraction (mediated by phonons) between the electrons. In some metals, this attractive force is greater than the Coulomb repulsion near the

Fermi surface, and the electrons can form bound states called *Cooper pairs* (as mentioned in the last section). In a superconductor we have to take into account the interactions between all the electrons at the same time. The electrons near the Fermi surface form Cooper pairs with the same total momentum. If the momentum is different from zero, there is a net current. Since all the electrons are paired together, we cannot slow down one single electron; we have to do this to all electrons at the same time. This is not easy to do and we have the phenomenon of superconductivity. The electron–phonon interaction is the main ingredient of BCS theory, mentioned in the last section.

Fermi liquid theory however does not describe strongly correlated systems, including the normal state of high temperature superconductors and heavy fermion compounds, the so called 'non-Fermi liquids' or strange metals [31]. For a non-Fermi liquid we do not know the structure of the spectral function. We can however define a Fermi surface as the surface in momentum space where there are gapless excitations, which in turn should result in the non-analytic behavior of the spectral function at $k = k_F$ and $\omega = 0$.

As we have mentioned before, poles of the fermion Green's function signal the existence of quasi-particles and in a normal Fermi liquid the excitations are sharp and long lived. In a non-Fermi liquid there are no stable quasi-particles. In these systems the single-particle lifetime is short and as a result, such excitations can no longer be treated as a quasi-particle. Thus the degrees of freedom are not long-lived quasi-particles. This shows anomalous behavior, such as a linear dependence of the resistivity with temperature, while a Fermi liquid has a resistivity that behaves as T^2 [32].

Experimental results from angle resolved photoemission spectroscopy for the strange metal region in high T_C cuprates show that the data can be fitted to the following expression [33]:

$$G^R(k, \omega) = \frac{h}{\omega - v_F(k - k_F) + \Sigma(k, \omega)}, \qquad (7.25)$$

with $\Sigma(\omega) \approx c\omega \ln \omega + d\omega$, where c is real and d is complex.

A proper theoretical framework from which to treat these systems is lacking. Numerical methods have the well-known 'fermion sign problem'. It is believed that AdS/CFT correspondence could give some insight into this subject. However this is not an easy task, since we do not have a complete understanding of the behavior of strongly correlated metals using standard approaches, and therefore we should be cautious in the interpretation of the results from the gravity duality. Here I will present a brief sketch of the formalism involved, following mainly reference [34], where the author examined the dynamics of fermions, hoping that his description could capture some universal features of strongly interacting fermions. I have followed this reference mainly for pedagogical reasons, as it is relatively simple. The reader should consult [27, 31, 33–43] for more detailed calculations and discussions. What I will do here is sufficient for the reader to get an idea of the formalism. If calculations are intended, the reader is advised to study the Dirac equation a little further.

The action studied here will be similar to the one studied in the last section, replacing the Klein–Gordon term by a Dirac term:

$$S = \frac{1}{k_4^2} \int d^4x \sqrt{g} \left[\frac{R}{4} - \frac{1}{4} F_{\mu\nu} F^{\mu\nu} + \frac{3}{2} \right]$$

$$+ \int d^4x \sqrt{g} \left[\overline{\psi} \gamma^\mu D_\mu \psi - m \overline{\psi} \psi \right] + \int d^3x \sqrt{g_\varepsilon} \, \overline{\psi} \psi. \tag{7.26}$$

In this section $D_\mu = \partial_\mu + \frac{1}{2} \omega_\mu^{bc} \Sigma_{bc} - iA_\mu$ is the covariant derivative, ω_μ^{bc} are the spin connections and $\Sigma_{bc} = \frac{1}{4} [\Gamma_b, \Gamma_c]$, with Γ_a the gamma matrices $\gamma^\mu = e_a^\mu \Gamma^a$, where e_a^μ is the vielbein. Note that in curved space time the Dirac action is modified and contains both a vielbein and a spin connection, as was discussed in section 3.1. The vielbein defines a local rest frame, allowing the constant Dirac matrices to act at each space time point. The cosmological constant has been set to be -1. Ψ is a four-component Dirac spinor with $\overline{\psi} = \psi^\dagger i \Gamma^0$. This spinor is the source field which is coupled with a fermionic field in boundary theory. The field ψ is in duality with some fermionic composite operator O in boundary theory. The conformal dimension of O is related to the mass of ψ by: $\Delta = 3/2 + mL$ [43]. The last term in the action is a boundary term defined at $r = \varepsilon$ where g_ε is the absolute value of the determinant of the induced metric on the boundary. This term was added to make the variational principle well posed and to cancel any divergences.

The action (7.26) has the black hole solution studied in section 4.4, written here using a different notation (to agree with reference [34]):

$$ds^2 = \frac{1}{r^2} \left\{ \alpha^2 [-f^2(r) dt^2 + dx^2 + dy^2] + \frac{dr^2}{f^2(r)} \right\}, \tag{7.27}$$

$$A_t(r) = q\alpha(r - 1),$$

where $f(r) = \sqrt{1 + q^2 r^4 - (1 + q^2) r^3}$. In this coordinate system, the horizon of the black hole is at $r = 1$. The temperature of the black hole is

$$T = \frac{\alpha}{4\rho} (3 - q^2). \tag{7.28}$$

In the following we will take $m = 0$ for simplicity and work in the probe limit. As a consequence we can focus on the chiral mode. Assuming a plane wave solution for the left chiral modes ψ_- and $\overline{\psi}_-$ which satisfy $\Gamma^5 \psi_- = -\psi_-$ and $\overline{\psi}_- \Gamma^5 = -\overline{\psi}_-$, we write

$$\psi(t,x,y,r) = e^{-i(\omega t - k_x x - k_y y)} \psi_-(r) \qquad \overline{\psi}(t,x,y,r) = e^{i(\omega t - k_x x - k_y y)} \overline{\psi}_-(r). \tag{7.29}$$

The gamma matrices can be written in several equivalent sets. Here we use what is called chiral representation, where the matrices are given by

$$\Gamma^0 = \begin{pmatrix} 0 & -I \\ I & 0 \end{pmatrix} \qquad \Gamma^i = \begin{pmatrix} 0 & \sigma^i \\ \sigma^i & 0 \end{pmatrix}, \tag{7.30}$$

where σ^i are the Pauli matrices, and I denotes the 2×2 unit matrix. The equation of motion can be written as

$$\left[\frac{i\omega r}{\alpha f} + \frac{iqr(r-1)}{f} + \frac{ik \cdot \sigma}{\alpha} r + \left(\frac{rf'}{2} - \frac{3f}{2} + rf\partial_r \right)\sigma^z \right] \psi_-(r) = 0,$$

$$\overline{\psi}_-(r) \left[\frac{i\omega r}{\alpha f} + \frac{iqr(r-1)}{f} - \frac{ik \cdot \sigma}{\alpha} r + \left(\frac{rf'}{2} - \frac{3f}{2} + \overleftarrow{\partial}_r rf \right)\sigma^z \right] = 0, \qquad (7.31)$$

where the prime denotes a derivative with respect to r. At zero temperature, imposing the ingoing boundary conditions, we can write near the horizon ($r \to 1$)

$$\psi_-(r) \approx e^{i\omega \frac{1}{6\alpha(1-r)}} \begin{pmatrix} 0 \\ 1 \end{pmatrix} \qquad \overline{\psi}_-^T(r) \approx e^{-i\omega \frac{1}{6\alpha(1-r)}} \begin{pmatrix} 1 \\ 0 \end{pmatrix}. \qquad (7.32)$$

Near the boundary ($r \to 0$), the spinors behave as $\psi_-(r) \approx r^{3/2}\chi$ and $\overline{\psi}_-(r) \approx r^{3/2}\overline{\chi}$, where χ and $\overline{\chi}$ are chosen so that the solutions satisfy the ingoing boundary conditions near the horizon. Near the boundary, the solutions can be written as

$$\psi_-(r) \approx \left(\frac{r}{\alpha} \right)^{3/2} \begin{pmatrix} P(\omega, \mathbf{k}) \\ 1 \end{pmatrix} \eta \qquad \overline{\psi}_-^T(r) \approx \left(\frac{r}{\alpha} \right)^{3/2} \begin{pmatrix} Q(\omega, \mathbf{k}) \\ 1 \end{pmatrix} \eta^*, \qquad (7.33)$$

where η and η^* are Grassmann numbers used to impose boundary conditions. (Grassmann numbers anticommute with each other but commute with ordinary numbers.)

The spectral function is given by $A(\omega, \mathbf{k}) = \lim_{\delta \to 0} \Im G^R(\omega + i\delta, \mathbf{k})$, where the Green's function in the boundary theory can be obtained from the action (7.26) evaluated at the saddle point. The bulk contribution vanishes there and only the boundary term contributes [34]. We have then

$$G^R(\omega, \mathbf{k}) = i[P(\omega, \mathbf{k})Q(\omega, \mathbf{k}) + 1]. \qquad (7.34)$$

Lee [34] solved the equation of motion numerically to obtain $A(\omega, \mathbf{k})$. He did not find a quasi-particle peak, implying that the fermions were in a non-Fermi liquid state, but the Fermi surface was still well defined. Different choices of gamma matrices may lead to different AdS/CFT prescriptions for the Green's function.

Liu et al [39] discussed the case $m = 0$ in detail. The case $m \neq 0$ was considered by Cubrovic et al [42] in a very detailed calculation. They presented evidence that the AdS dual description of strongly coupled field theories can describe the emergence of a Fermi liquid from a quantum critical state. They computed the spectral function of fermions in field theory. By increasing the fermion density away from the relativistic quantum critical point, a state emerged with all the features of the Fermi liquid. Tuning the scaling dimensions of the critical fermion field they found that the quasi-particles disappeared at a QPT of a purely statistical nature.

Brynjolfsson et al [38] reproduced qualitatively the anomalous specific heat of heavy fermion metals using gravitational dual models with dynamical critical scaling $z > 1$.

Faulkner *et al* [43] discovered a class of non-Fermi liquids using gauge gravity duality, where the single-particle spectral function and transport behavior resembled those of strange metals, in particular the resistivity was proportional to the temperature and the self-energy $\Sigma(\omega)$ was of the form of (7.25). Liu *et al* [39] considered an AdS–Einstein–Maxwell theory with a charged massive scalar and a charged massive fermion. They found that depending on the relative conformal weight, the ground state was either a pure AdS Reissner–Nordström black hole, a holographic superconductor, an electron star, or a novel mixed state that was best characterized as a hairy electron star.

7.3 Entanglement

Besides correlation functions and the thermodynamic potential, other observations of interest can be calculated using the correspondence principle. One of these is the entanglement entropy. Entanglement provides a measure of the amount of quantum non-locality in a quantum state. Entanglement entropy (EE) is a measure of how much a given quantum state is entangled [40, 44]. It is particularly useful in condensed matter as a tool to characterize quantum critical points or topological orders.

For a pure state the density matrix is given by

$$\rho_{\text{tot}} = |\psi\rangle\langle\psi|, \qquad (7.35)$$

where $|\psi\rangle$ is the wave function of the system. For a mixed state in a generic quantum system at finite temperature, we have for the canonical ensemble

$$\rho_{\text{tot}} = e^{-\beta H}/Z, \qquad (7.36)$$

where $Z = Tr e^{-\beta H}$.

Now we divide the system into two parts A and B and use the entropy as a measure of the correlation between the subsystems. The reduced density matrix for the subsystem A is given by integrating out the degrees of freedom in the subsystem B, that is $\rho_A = Tr_B \rho_{\text{tot}}$. The entanglement entropy S_A is defined by

$$S_A = -Tr_A[\rho_A \ln \rho_A]. \qquad (7.37)$$

The standard thermal entropy is therefore obtained as a particular case of the EE, i.e. A = total space.

The holographic method reduces the calculation of the EE to the evaluation of minimal surfaces [45]. The entropy in a region of space time is given by the area of the surface that bounds it (we should remember that the holographic principle was born from the fact that the degrees of freedom in gravity are proportional to the area instead of the volume).

Ryu and Takayanagi [40] used the following procedure to calculate the EE. The theory for which they intended to calculate the EE was applied to the boundary of the AdS space and a division into regions A and B was carried out on this boundary. A closed surface was then extended from region A into the bulk. The calculation of the EE is equivalent to finding a surface with a minimum area (see figure 7.3).

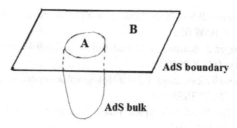

Figure 7.3. Calculation of the entanglement entropy for the region A.

A large class of systems related to entanglement entropy in Fermi systems was studied using the following metric [44, 46]:

$$ds^2 = r^{\frac{2\theta}{d-1}}\left(-\frac{dt^2}{r^{2z}} + \frac{dr^2 + dx_i^2}{r^2}\right). \tag{7.38}$$

Under the transformation $t \rightarrow \lambda^z t$, $x_i \rightarrow \lambda x_i$, $r \rightarrow \lambda r$, the metric transforms as

$$ds \rightarrow \lambda^{\frac{\theta}{d-1}}ds, \tag{7.39}$$

where z is the dynamical critical exponent and θ is the violation of the hyperscaling exponent.

References

[1] Bardeen J, Cooper L N and Schrieffer J R 1957 Microscopic theory of superconductivity *Phys. Rev.* **106** 162

[2] Berdnoz J G and Mueller K A 1986 Possible high T_C superconductivity in La-La-Cu-O systems *Z. Phys.* B **64** 189

[3] Kamihara Y, Watanabe T, Hirano M and Hosono H 2008 Iron based layered super-conductor La[O$_{1-x}$F$_x$]FeAs ($x = 0.05-0.12$) with $T_C = 26$ K *J. Am. Chem. Soc.* **130** 3296

[4] Hartnoll S A Quantum critical dynamics from black holes arXiv:0909.3553 [cond-mat.str-el]

[5] Herzog C P 2009 Lectures on holographic superfluidity and superconductivity *J. Phys. A: Math. Gen.* **42** 343001

[6] Hartnoll S A, Herzog C P and Horowitz G T 2008 Building a holographic superconductor *Phys. Rev. Lett.* **101** 031601

[7] Hartnoll S A, Herzog C P and Horowitz G T Holographic superconductors arXiv:0810.1563 [hep-th]

[8] Horowitz G T and Roberts M M 2008 Holographic superconductors with various condensates *Phys. Rev.* D **78** 126008

Horowitz G T 2011 Introduction to holographic superconductors *Lecture Notes in Physics* **828** 313

Horowitz G T and Roberts M M 2009 Zero temperature limit of holographic superconductors *JHEP* **911** 15

[9] Maeda K and Okamura T 2008 Characteristic length of an AdS/CFT superconductor *Phys. Rev.* D **78** 106006

[10] Gubser S S and Pufu S S 2008 The gravity dual of a p-wave superconductor *JHEP* **0811** 033

[11] Roberts M M and Hartnoll S A 2008 Pseudogap and time reversal breaking in a holographic superconductor *JHEP* **0808** 035

[12] Ammon M, Erdmenger J, Kaminski M and Kerner P 2009 Flavor superconductivity from gauge/gravity duality *JHEP* **0910** 067

[13] Nishioka T, Ryu S and Takayanaga T 2010 Holographic superconductor/insulator transition at zero temperature *JHEP* **1003** 131

[14] Horowitz G T and Way B 2010 Complete phase diagram for a holographic superconductor/ insulator system *JHEP* **1011** 011

[15] Herzog C P 2010 Analytic holographic superconductor *Phys. Rev.* D **81** 126009

[16] Salvio A 2012 Holographic superfluids and superconductors in dilaton-gravity *JHEP* **1209** 124

[17] Domenech O, Montull M, Pomarol A, Salvio A and Silva P J Emergent gauge fields in holographic superconductors *JHEP* **1008** 920100 033

[18] Ge X H, Wang B, Wu S F and Yang G H 2010 Analytical study on holographic super-conductors in external magnetic field *JHEP* **1008** 181601

[19] Arean D, Bertolini M, Evslin J and Prochzka T On holographic superconductors with DC currents *JHEP* arXiv:1003.5661 [hep-th]

[20] Montull M, Pujolas O, Salvio A and Silva P J 2012 Magnetic response in the holographic insulator/superconductor transition *JHEP* **1204** 135

[21] Salvio A 2013 Superconductors, superfluidity and holography *J. Phys.: Conf. Ser.* **442** 012040

[22] Salvio A Transitions in dilaton holography with global or local symmetries *JHEP* arXiv:1302.4898 [hep-th]

[23] Iqbal N, Liu H, Mezei M and Si Q 2010 Quantum phase transitions in holographic models of magnetism and superconductors *Phys. Rev.* D **82** 045002

[24] Wenger T Holographic superconductivity http://studentarbeten.chalmers.se/publication/ 162268-holographic-superconductivity

[25] Hartnoll S A 2009 Lectures on holographic methods for condensed matter physics *Class. Quantum Grav.* **26** 224002

[26] Horowitz G T 2011 Surprising connections between general relativity and condensed matter *Class. Quantum Grav.* **28** 114008

[27] Zaanen J, Sun Y W, Liu Y and Schalm K The AdS/CFT manual for plumbers and elec-tricians www.lorentz.leidenuniv.nl/~kschalm/papers/adscmtreview.pdf

[28] Phillips P 2003 *Advanced Solid State Physics* (Boulder, CO: Westview Press)

[29] Hartnoll S A and Tavanfar A 2011 Electron stars for holographic metallic criticality *Phys. Rev.* D **83** 046003
Cubrovic M, Liu Y, Schalm K, Sun Y W and Zaanen J Spectral probes of the holographic Fermi groundstates: dialing between the electron star and AdS Dirac hair arXiv:1106.1798 [hep-th]

[30] Iqbal N, Liu H and Mezei M Lectures on holographic non-Fermi liquids and quantum phase transitions arXiv:1110.3814 [hep-th]

[31] Stewart G R Non-Fermi-liquid behavior in d- and f-electron metals *Rev. Mod. Phys.* **66** 92001 797

[32] Sachdev S 2011 Strange metals and AdS/CFT correspondence *J. Stat. Mech.* **1011** 11022

[33] Abrahams E and Varma C M 200 What angle-resolved photoemission experiments tell about the microscopic theory for high-temperature superconductors *Proc. Natl. Acad. Sci. USA* **97** 5714

[34] Lee S S 2009 A non-Fermi liquid from a charged black hole: a critical Fermi ball *Phys. Rev. D* **79** 086006

[35] Liu H, McGreevy J and Vegh D 2011 Non-Fermi liquids from holography *Phys. Rev. D* **83** 065029

[36] Medvedyeva M V, Gubankowa E, Cubrovik M, Schalm K and Zaanen J Quantum corrected phase diagram of holographic fermions arXiv:1302.5149 [hep-th]

[37] Sachdev S 2011 A model of a Fermi liquid using gauge-gravity duality *Phys. Rev. D* **84** 066009

Huijse L and Sachdev S 2011 Fermi surfaces and gauge-gravity duality *Phys. Rev. D* **84** 026001

Sachdev S 2010 Holographic metals and the fractionalized Fermi liquid *Phys. Rev. Lett.* **105** 151602

[38] Brynjolfsson E J, Danielsson U H, Thorlacius L and Zingg T Black hole thermodynamics and heavy fermion metals arXiv:1003.5361 [hep-th]

[39] Liu Y, Schalm K, Sun Y W and Zaanen J Bose–Fermi competition in holographic metals arXiv:1307.4572 [hep-th]

[40] Ryu S and Takayanagi T 2006 Holographic derivation of entanglement entropy from AdS/CFT *Phys. Rev. Lett.* **96** 181602

[41] Faulkner T, Liu H, McGreevy J and Vegh D Emergent quantum criticality, Fermi surface, and AdS$_2$ arXiv:0907.2694 [hep-th]

[42] Cubrovic M, Zaanen J and Schalm K 2009 String theory, quantum phase transitions and the emergent Fermi-liquid *Science* **325** 439

Cubrovic M, Zaanen J and Schalm K Fermions and the AdS/CFT correspondence: quantum phase transitions and the emergent Fermi-liquid arXiv:0904.1993 [hep-th]

[43] Faulkner T, Iqbal N, Liu H, McGreevy J and Vegh D From black holes to strange metals arXiv:1003.1728 [hep-th]

[44] Nishioka T, Ryu S and Takayanagi T 2009 Holographic entanglement entropy: an overview *J. Phys. A: Math. Gen.* **42** 504008

[45] Green A G An introduction to gauge gravity duality and its application in condensed matter arXiv:1304.5908 [cond-mat]

[46] Ogawa N, Takayanagi T and Ugajin T 1012 Holographic Fermi surfaces and entanglement entropy *JHEP* **1201** 125

Chapter 8

Conclusions

The procedure to use AdS/CFT correspondence can be sumarized as:

1. We choose a metric allowing for black holes and asymptotically anti-de Sitter space. Applying the boundary theory at finite temperature and finite density corresponds to putting a black hole in the bulk space.
2. We write an action in the bulk for one, or more, fields. The full action may be complicated with many fields. However, it is generally possible to use a small number of fields that capture the physics of interest.
3. In the large N limit of boundary theory (strong coupling limit) we solve the classical field equations in the bulk. Since they are second order differential equations, the solutions can be found numerically.
4. We find the asymptotics of the fields at the boundary. The ingoing amplitude at the boundary represents the source, the outgoing amplitude gives the response for the theory at the boundary. We can then calculate the Green's function as was explained in chapter 5. For every field in the bulk, there is a gauge invariant operator in the boundary theory.

Several generalizations are possible. For instance, adding a term

$$\int d^{d+1}\sqrt{g}\,\frac{\gamma L^2}{e^2}C_{\mu\eta\sigma\rho}F^{\mu\nu}F^{\sigma\rho},\qquad(8.1)$$

where γ is a parameter and $C_{\mu\eta\sigma\rho}$ is the Weyl curvature tensor, a broader class of problems (including the Bose–Hubbard model) can be studied [1]. The Weyl tensor is basically the Riemann tensor with all of its contractions removed. In n dimensions we have

$$C_{\rho\sigma\mu\nu} = R_{\rho\sigma\mu\nu} - \frac{2}{(n-2)}\left(g_{\rho[\mu}R_{\nu]\sigma} - g_{\sigma[\mu}R_{\nu]\rho}\right) + \frac{2}{(n-1)(n-2)}g_{\rho[\mu}g_{\nu]\sigma}R\qquad(8.2)$$

Here [...] means antisymmetrization, where in the sum over permutations of indices an odd number of exchanges are given a minus sign.

Another interesting extension [2, 3] is provided by adding a real bulk scalar field (called a *dilaton*) which couples to the gauge field by changing the gauge coupling and leading to the action [4]

$$S = \int d^4x \left(R - 2\Lambda - \frac{1}{2}\left(\partial_\mu \varphi\right)^2 - \frac{Z(\varphi)}{4}F_{\mu\nu}F^{\mu\nu} + V(\varphi) \right) \tag{8.3}$$

The presence of φ can allow a charged asymptotically AdS black hole to have vanishing entropy in the low temperature limit. A Salvio [2, 3] has used the dilaton-gravity system to study superconductors and superfluids.

We remark that the formalism only allows us to calculate correlation functions for operators of field theory. It does not give us a Lagrangian. The gravity description captures the macroscopic dynamics of the order parameter, but does not explain its microscopic origin. Different values for the coupling constants in the bulk will correspond to different theories in the boundary. At present we cannot deduce the gravitational dual corresponding to any particular microscopic condensed matter model. As pointed out by Hartnoll [5], in the immediate future it is unlikely that values for experimental quantities obtained holographically could aspire to be more than useful benchmarks. However a promising aspect of the formalism is that it provides examples of theories without a quasi-particle description where computations can be performed. Questions for experiments that would never be asked to depart from conventional theories can come to mind. Using this formalism, charged bosonic operators have been found to lead to superconducting phases while charged fermionic operators capture non-Fermi liquid behavior. Herzog [6] has pointed out that high T_C superconductors may not be good candidates for holographic techniques; heavy fermions or some other condensed matter systems seem to be more suited to this approach. Some examples of emergent $U(1)$ fields that are well understood and relevant for condensed matter physics are given in [7].

Metallic quantum critical points seem to be generically unstable to the formation of new phases and, as pointed out by Green [8], when these points are approached sufficiently close in temperature and tuning parameters, new physics (leading to a new quantum order) intervenes before criticality. Figure 8.1 shows a sketch of a phase diagram for these systems. It is believed that there is a quantum critical point p hidden below the dome. This behavior is not well understood using standard techniques. There is a hope that the holographic principle could give us some insight into what happens here.

There are several other applications not considered here. For instance, the duality principle can be used to gain new perspectives and insights into the out of equilibrium behavior of quantum systems [8]. Horowitz *et al* [9] presented a gravitational dual of a system with a lattice by introducing a neutral scalar field with boundary conditions corresponding to a periodic source. The study of non-relativistic holographic systems that realize scale symmetry without conformal symmetry (i.e. $z > 1$) has progressed in recent years and such models have been

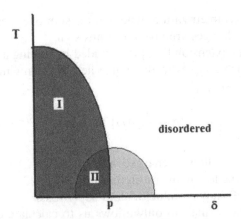

Figure 8.1. Sketch of a quantum critical phase diagram (adapted from [8]). I: ordered phase, II: new quantum phase.

applied to condensed matter systems [10]. There have been some attempts to obtain a constructive derivation of gauge gravity duality that does not involve the machinery of string theory. One of these attempts is the multi-scale entanglement renormalization ansatz (MERA) [11].

The holographic method has introduced a new set of mathematical tools and new insights can arise in the future. Several physicists working in the subject are optimistic.

References

[1] Huijse L, Sachdev S and Swingle B 2012 Hidden Fermi surfaces in compressible states of gauge-gravity duality *Phys. Rev.* B **85** 035121

[2] Salvio A 2012 Holographic superfluids and superconductors in dilaton-gravity *JHEP* **1209** 124

[3] Salvio A 2013 Superconductors, superfluidity and holography *J. Phys.: Conf. Ser.* **442** 012040

[4] Zaanen J, Sun Y W, Liu Y and Schalm K The AdS/CFT manual for plumbers and electricians www.lorentz.leidenuniv.nl/~kschalm/papers/adscmtreview.pdf

[5] Hartnoll S A 2009 Lectures on holographic methods for condensed matter physics *Class. Quantum Grav.* **26** 224002

[6] Herzog C P 2009 Lectures on holographic superfluidity and superconductivity *J. Phys. A: Math. Gen.* **42** 343001

[7] Sachdev S Exotic phases and quantum phase transitions: model systems and experiments arXiv:09014103 [cond-mat.str-el]

[8] Green A G An introduction to gauge gravity duality and its application in condensed matter arXiv:1304.5908 [cond-mat]

[9] Horowitz G, Santos J E and Tong D Optical conductivity with holographic lattices arXiv:1204.0519 [hep-th]

[10] Keranen V and Thorlacius L Holographic geometries for condensed matter applications arXiv:1307.2882 [gr-qc]

[11] Swingle B 2012 Entanglement renormalization and entropy *Phys. Rev* D **86** 065007

AdS/CFT Correspondence in Condensed Matter

Antonio Sergio Teixeira Pires

Appendix A

The Schwarzschild solution

In 1915 Karl Schwarzschild found the first exact solution for Einstein's equation in empty space. To obtain this solution we look for a spherically symmetric and time independent solution to Einstein's equations in empty space and the zero cosmological constant. We start with the ansatz metric in spherical coordinates (t, r, θ, ϕ)

$$ds^2 = -e^{2\alpha(r)}dt^2 + e^{2\beta(r)}dr^2 + r^2 d\Omega^2, \tag{A.1}$$

where $d\Omega^2 = d\theta^2 + \sin^2\theta d\phi^2$. We can write an r-dependent term by multiplying the last term in (A.1), but this term can be eliminated by a change of radial coordinate. See reference [1] for more details about this choice. Using equation (3.6) we evaluate the Christoffel symbols

$$
\begin{aligned}
&\Gamma^t_{tr} = \partial_r\alpha && \Gamma^r_{tt} = e^{2(\alpha-\beta)}\partial_r\alpha && \Gamma^r_{rr} = \partial_r\beta, \\
&\Gamma^\theta_{r\theta} = \frac{1}{r} && \Gamma^r_{\theta\theta} = -re^{-2\beta} && \Gamma^\phi_{r\phi} = \frac{1}{r}, \\
&\Gamma^r_{\phi\phi} = -re^{-2\beta}\sin^2\theta && \Gamma^\phi_{\theta\phi} = \frac{\cos\theta}{\sin\theta} && \Gamma^\theta_{\phi\phi} = -\sin\theta\cos\theta.
\end{aligned}
\tag{A.2}
$$

The other terms vanish or are related to the above expression by symmetry. From equation (3.7) we obtain the Riemann tensor

$$
\begin{aligned}
&R^t_{rtr} = \partial_r\alpha\partial_r\beta - \partial_r^2\alpha - (\partial_r\alpha)^2 && R^t_{\theta t\theta} = -re^{-2\beta}\partial_r\alpha, \\
&R^t_{\phi t\phi} = -re^{-2\beta}\sin^2\theta\partial_r\alpha && R^r_{\theta r\theta} = re^{-2\beta}\partial_r\beta, \\
&R^r_{\phi r\phi} = re^{-2\beta}\sin^2\theta\partial_r\beta && R^\theta_{\phi\theta\phi} = (1 - e^{-2\beta})\sin^2\theta.
\end{aligned}
\tag{A.3}
$$

doi:10.1088/978-1-627-05309-9ch9

The Ricci tensor is obtained by contraction using equation (3.9)

$$R_{tt} = e^{2(\alpha-\beta)}\left[\partial_r^2\alpha + (\partial_r\alpha)^2 - \partial_r\alpha\partial_r\beta + \frac{2}{r}\partial_r\alpha\right]$$

$$R_{rr} = -\partial_r^2\alpha - (\partial_r\alpha)^2 + \partial_r\alpha\partial_r\beta + \frac{2}{r}\partial_r\beta \tag{A.4}$$

$$R_{\theta\theta} = e^{-2\beta}\left[r(\partial_r\beta - \partial_r\alpha) - 1\right] + 1, \quad R_{\phi\phi} = \sin^2\theta R_{\theta\theta}.$$

The Einstein equation in empty space is given by $R_{\mu\nu} = 0$. The components R_{tt} and R_{rr} vanish independently, which allows us to write

$$e^{2(\beta-\alpha)}R_{tt} + R_{rr} = \frac{2}{r}\partial_r(\alpha + \beta) = 0, \tag{A.5}$$

which gives $\alpha + \beta = c$, where c is a constant. Rescaling the time coordinate by $t \to e^{-c}t$ we can eliminate c. Thus we find $\alpha = -\beta$. The equation $R_{\theta\theta} = 0$ gives

$$e^{2\alpha}(2r\partial_r\alpha + 1) = 1 \quad \text{or} \quad \partial_r(re^{2\alpha}) = 1. \tag{A.6}$$

The solution of this equation is given by

$$e^{2\alpha} = 1 - \frac{R_S}{r}, \tag{A.7}$$

where R_S is an integration constant. The metric can then be written as

$$ds^2 = -\left(1 - \frac{R_S}{r}\right)dt^2 + \left(1 - \frac{R_S}{r}\right)^{-1}dr^2 + r^2(d\theta^2 + \sin^2\theta d\varphi^2). \tag{A.8}$$

Calculating the metric in the weak field limit (Newtonian limit) we obtain $g_{tt} \approx 1 - 2G_N m/r$, which allows us to write $R_S = 2G_N m$, where G_N is the Newton constant and m is a parameter which we identify with the Newtonian mass. It can be shown that the Schwarzschild metric is the unique vacuum solution with spherical symmetry for the Einstein equation. The parameter R_S is called the Schwarzschild radius or horizon. For more details see reference [1].

It may be strange that we started with empty space and finished with a solution with mass at the origin. The same thing happens in the Newtonian gravitation case. To see what I mean let us consider Newton's equation in empty space

$$\nabla^2\phi = 0, \tag{A.9}$$

where ϕ is the Newtonian potential. For a spherically symmetric solution, equation (A.9) becomes

$$\frac{1}{r^2}\frac{\partial}{\partial r}\left(r^2\frac{\partial\phi}{\partial r}\right) = 0, \tag{A.10}$$

with the solution

$$\phi = -\frac{c_1}{r} + c_2. \tag{A.11}$$

This solution can be interpreted as the potential for a mass m at the origin $r = 0$. Thus, it is not surprising that the Schwarzschild solution for the empty space can describe a mass at the origin.

The main interest of the Schwarzschild solution is to describe the orbits around a star or planet. Note that as $m \to 0$, or $r \to \infty$ we recover Minkowski space, which is to be expected. A remarkable solution describing new phenomena of interest happens for $r < R_S$. In this region time and radial direction will change with respect to each other and nothing can escape from inside. However, for the standard bodies of our Universe, stars and planets, this solution is of no consequence since R_S is smaller than the radius of the object (and therefore in a region where the solution is not valid). For our Sun, for instance, R_S is of the order of three kilometers. Only more recently, the idea that the metric (A.8) could describe a real object (that was named a *black hole* by Wheeler in 1967) was taken seriously. I would expect the reader to know what a black hole is and I am not going into further detail here. It is interesting to remark that there is no non-singular Schwarzschild-like solution in (2 + 1) dimensions for Einstein equations in empty space and for a zero cosmological constant.

Unlike the flat space time case (or the de Sitter space), black holes in AdS space are more complicated [2]. Besides their spherical topology, they can have toroidal or hyperbolic topologies. The simplest example of an asymptotically AdS black hole in three spatial dimensions is given by

$$ds^2 = -\left(1 - \frac{\mu}{r} + \frac{r^2}{L^2}\right)dt^2 + \left(1 - \frac{\mu}{r} + \frac{r^2}{L^2}\right)^{-1}dr^2 + r^2 d\Omega^2, \quad (A.12)$$

where μ is proportional to the mass and L is the radius of curvature of the AdS space.

Hawking and Bekenstein found that the entropy of a black hole is proportional to the horizon area. We have the relation

$$S_{BH} = \frac{A}{4} = \frac{k_B}{c\hbar}4\pi M^2, \quad (A.13)$$

where S_{BH} is the thermal entropy associated with a surface having the area of the horizon, divided into cells with a size set by the Planck length, where every cell contains one bit of information [1].

References

[1] Carrol S M 2004 *Spacetime and Geometry* (Reading, MA: Addison-Wesley)
[2] Witten E 1998 Anti-de Sitter space, thermal phase transition, and confinement in gauge theories *Adv. Theor. Math. Phys.* **2** 505

Appendix B

Euler–Lagrange equations

Here I show how to derive the equations of motion from the action. Let us consider a field theory with a Lagrangian density L. The action is given by

$$S = \int \sqrt{g} L\left(\phi, \nabla_\mu \phi\right) \mathrm{d}^n x. \tag{B.1}$$

The Euler–Lagrange equation obtained from (B.1) is

$$\frac{\partial L}{\partial \phi} - \nabla_\mu \left(\frac{\partial L}{\partial \left(\nabla_\mu \phi \right)} \right) = 0. \tag{B.2}$$

For example, if the action is given by

$$S_\phi = \int \left[-\frac{1}{2} g^{\mu\nu} \left(\nabla_\mu \phi \right)(\nabla_\nu \phi) - V(\phi) \right] \sqrt{g}\, \mathrm{d}^n x, \tag{B.3}$$

the equation of motion is

$$g^{\mu\nu} \nabla_\mu \nabla_\nu \phi - \frac{\mathrm{d}V}{\mathrm{d}\phi} = 0 \tag{B.4}$$

or

$$\frac{1}{\sqrt{g}} \partial_\mu \left(\sqrt{g}\, g^{\mu\nu} \partial_\nu \phi \right) - \frac{\mathrm{d}V}{\mathrm{d}\phi} = 0. \tag{B.5}$$

As we saw in section 3.1, the action S_H should be added to S_ϕ to give the total action [1]

$$S = \frac{S_H}{16\pi G} + S_\phi. \tag{B.6}$$

To obtain Einstein's equation we also have to vary S_ϕ with respect, not to ϕ, but to the inverse metric. Using equation (3.15) (which is the contribution of the first term in B.6) we arrive at the following equation:

$$R_{\mu\nu} - \frac{1}{2} R g^{\mu\nu} + \Lambda g_{\mu\nu} = 8\pi G T_{\mu\nu}, \qquad (B.7)$$

where the energy-momentum tensor $T_{\mu\nu}$ is given by

$$T_{\mu\nu} = \nabla_\mu \phi \nabla_\nu \phi - \frac{1}{2} g_{\mu\nu} g^{\rho\sigma} \nabla_\rho \phi \nabla_\sigma \phi - g_{\mu\nu} V(\phi). \qquad (B.8)$$

The matter field changes the metric of the empty space (we say that it back reacts) and we have to solve the coupled equations (B.4) and (B.7) to find $g_{\mu\nu}$ and ϕ.

At a first approximation we can neglect the effect of back reaction and use the metric for empty space in equation (B.4), as was done in section 4.2.

Reference

[1] Carrol S M 2004 *Spacetime and Geometry* (Reading, MA: Addison-Wesley)